THE SPECIAL THEORY OF RELATIVITY

THE SPECIAL THEORY OF RELATIVITY

H. Muirhead

University of Liverpool

A HALSTED PRESS BOOK

JOHN WILEY & SONS

NEW YORK – TORONTO

© H. Muirhead 1973

Published in the United Kingdom 1973 by
The Macmillan Press Ltd

Published in the U.S.A. and Canada by
Halsted Press, a Division of John Wiley & Sons, Inc., New York

Printed in Great Britain

Library of Congress Cataloging in Publication Data

Muirhead, Hugh
      The special theory of relativity

      "A Halsted Press Book."
      1.   Relativity (Physics)   2.   Lorentz transformations
3.   Symmetry (Physics)   I.   Title
QC6.M86   530.1'1   72–6874
ISBN 0–470–62357–8

Preface

It has always been a source of surprise to me that no undergraduate textbooks on special relativity have been written by high-energy nuclear physicists. They are the people who most commonly use relativity, and at the same time their approach has more in common with Einstein's original work than most textbooks now show. The latter place too much emphasis (in the opinion of the present author) on what one might call the 'clocks and rods' aspect and too little on the principle of the invariance of physical laws in different reference frames.

This book is an attempt to redress the balance. I have tried to set out the exploitation of the principle of Lorentz invariance in the way in which it is done in high-energy physics, and in doing so have tried to indicate that the basic concept of the invariance of physical laws is a very general one and goes far beyond the context of relativity. Furthermore the techniques developed by high-energy physicists for the construction of physical equations by invariance arguments are an important extension of the principles of dimensional analysis and deserve to be more widely known than they are at present.

In writing a textbook on special relativity from the high-energy physicist's point of view, I have also felt the necessity to include material on spin and polarisation in relativistic situations. This is a subject of considerable importance in high-energy physics, and is readily amenable to presentation to undergraduates. Nevertheless, the topic is ignored by most writers on special relativity.

The basic plan of the book is that chapter 1 is concerned with historical matters. Chapter 2 deals with the basic Lorentz transformation, and the concept of the invariance of the interval between two events. This invariance principle, together with the implications of the rotation of space–time axes are examined in chapter 3. Chapter 4 is concerned with four-vectors and tensors, and in particular with the momentum four-vector. The treatment of spin and polarisation is made in chapter 5, and a further discussion of the invariance principle is made. Chapter 6 is concerned with dynamics, chiefly electrodynamics, and the invariance of the transformation properties of physical equations is exploited to examine the behaviour of the motion and polarisation vector ('spin') of charged particles in electrodynamic fields.

In introducing the Lorentz transformation, I was tempted of ignore the early work and worries of the turn of the century and to treat the subject axiomatically. However, I think it is good for students to know that physics rarely evolves in a neat orderly way and that a successful theory normally emerges only after a number of pathways have been tried and found to be blind alleys. For me one of the most interesting facts which emerged in writing chapter 1 was the realisation that Maxwell obviously had worries about the ether concept, and that they lead to a curiously defensive tone in his presentation of certain parts of his *Treatise on Electricity and Magnetism*.

Except in chapter 1, I have used MKS units for electromagnetic systems. They appear to me to be particularly inappropriate for a textbook on special relativity as they obscure rather than bring out the role of the velocity of light. However, as they are in general use I have reluctantly stuck to them.

I wish to thank I. G. Muirhead for reading the manuscript, and Drs J. Bailey and E. Picasso for supplying suitable diagrams for figures 6.4 and 6.5.

H. Muirhead

Contents

Acknowledgements

The author and publishers acknowledge with gratitude their indebtedness
to the following sources for permission to use copuright material.

	page
The Institution of Electrical Engineers for Abstract 2277 from *Physics Abstracts*, 1905.	1
G. D. Scott and M. R. Viner for their diagram of a plane grid moving at relativistic speeds, from the *American Journal of Physics*, **334**, 534, 1964.	46
R. S. Shankland for figure 7 from *Atomic and Nuclear Physics*, Macmillan and Company Limited, 1961.	57
The Royal Society for the table of Bradley's observations of γ-Draconis, from the *Philosophical Transactions of the Royal Society*, **35**, 637, 1728.	72
A. H. Compton and the American Physical Society for the graph showing the wavelength of scattered radiation as a function of the scattering angle, from the *Physical Review*, **21**, 483, 1923.	78

1

Historical notes

2277. *Electrodynamics of Moving Bodies.* **A. Einstein.** (Ann. d. Physik, **17.** 5. pp. 891–921, Sept. 26, 1905.)—A mathematical investigation divided into a kinematical and an electrodynamical part. The former treats (1) Definition of contemporaneousness (Gleichzeitigkeit), (2) The relativity of lengths and times, (3) Theory of space- and time-transformations from resting to uniformly translated systems, (4) Physical significance of the equations obtained for moving rigid bodies, &c., (5) Addition theorem for velocities. The second part deals with (6) Maxwell-Hertz equations, (7) Doppler's principle and aberration, (8) Transformation of energy of light, Theory of radiation pressure on a perfect reflector, (9) Transformation of Maxwell-Hertz equations with reference to convection currents, (10) Dynamics of slowly accelerated electrons.   E. H. B.

1.1   Galileo and Newton

The entry given above appears as abstract number 2277 in *Science Abstracts* *
in 1905. The paper which was abstracted is regarded as the final emergence of

---

* Section A, Physics, vol. 8, p. 713.

the theory of relativity in its presently accepted form. However like most great turning points in science this work did not suddenly emerge from nowhere, but was based on problems which had been known for many years and the attempts to solve these problems which had proved to be failures.

The principle of relativity first emerged in recognisable form in the work of Galileo and Newton. Galileo* argued that the vertical paths of falling objects did not allow one to conclude that the earth was stationary (and at the centre of the universe as proposed in the Ptolemaic system). He illustrated his point by giving an example of a rock dropped from the top of the mast of a ship. Regardless whether the ship is stationary or moving with constant velocity, the rock always lands at the foot of the mast and so the ship's state of motion cannot be deduced by an observer travelling with the ship and watching only this process. In a similar manner the earth's state of motion could not be deduced from rocks falling vertically.

Newton† was even more specific. In one of his corollaries to the laws of motion he states: 'The motions of bodies included in a given space are the same among themselves, whether that space is at rest, or moves uniformly forwards in a right line without any circular motion'. Notice that uniform motion is an important ingredient in both statements. The holding and drinking of an after dinner cup of coffee is a pleasant experience both in one's own (terrestrial) home or in a jet plane flying at 600 miles per hour, but not so if the latter hits a bumpy patch of air and sudden accelerations and decelerations are experienced. In most of this book we shall only consider uniform motion.

Let us see how the Newtonian concept works out in practice. We shall consider the collision of two billiard balls which are moving head-on towards each other.

In order to describe the motion of the balls before and after the collision in a quantitative manner we need a *reference or coordinate frame* in which the positions of the balls are measured as a function of time in order to determine their velocities. Consider the three-dimensional reference frame given in figure 1.1(a) and let the point P represent the position of one of the balls at

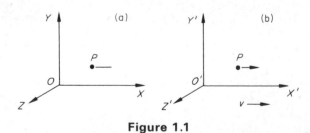

Figure 1.1

GALILEO *Dialogue Concerning the Two Chief World Systems—Ptolemaic and Copernican* (translated by S. Drake) University of California Press, 1953, p. 141.
† NEWTON *Mathematical Principles of Natural Philosophy (Principia)*, (p. 30 in translation by A. Motte, 1729). Reprinted by Dawson's of Pall Mall, 1968. The word 'right' in the quotation would nowadays be interpreted as 'straight'.

any time. For convenience we shall assume that the ball is moving parallel to the X axis. Then if the ball travels a distance dx in time dt the velocity is

$$u = \frac{\mathrm{d}x}{\mathrm{d}t} \tag{1.1.1}$$

and if no external forces are acting the velocity remains constant. Now in order to determine this quantity an observer must have been stationed in the XYZ frame equipped with an apparatus for measuring distances and times. Let us assume that we have a second observer stationed in another reference frame (figure 1.1(b)) which is moving past the XYZ frame with a constant velocity v in the X direction, and that he is similarly equipped. We assume that the point O' is at O at a time $t=0$ (zero), and that the observers synchronise their clocks at this time and that they remain synchronised for evermore. Then after a time t we find the position of P recorded in the XYZ frame as xyz and in the $X'Y'Z'$ frame as $x'y'z'$, where the coordinates are related by the equations

$$
\begin{aligned}
x' &= x - vt \\
y' &= y \\
z' &= z \\
t' &= t
\end{aligned}
\tag{1.1.2}
$$

These equations are collectively called the equations of the *Galilean transformation*. They can be extended to reference frames which are not necessarily moving parallel to the X axis, but we shall not bother with this refinement. From the equations (1.1.2) it is apparent that the velocity of P when measured by an observer in the $X'Y'Z'$ frame is given by

$$u' = \frac{\mathrm{d}x'}{\mathrm{d}t} = u - v \tag{1.1.3}$$

Now let us return to the problem of the collision of the two billiard balls. For simplicity we shall again assume that the balls travel parallel to the X axis. Let the balls have mass of m_1 and m_2 and velocities in the XYZ frame of u_1 and u_2 before the collision and U_1 and U_2 afterwards (figure 1.2).

The principle of conservation of momentum (which relates to Newton's

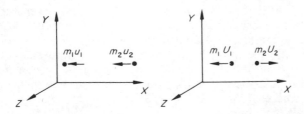

Figure 1.2

third law of motion) assumes the following form in the XYZ frame

$$m_1 u_1 + m_2 u_2 = m_1 U_1 + m_2 U_2 \tag{1.1.4}$$

(Strictly speaking the velocities and momenta are vectors but this property is irrelevant to our present discussion.)

Equation (1.1.3) then tells us that an observer in the $X'Y'Z'$ frame will record the process as

$$m_1 (u'_1 + v) + m_2 (u'_2 + v) = m_1 (U'_1 + v) + m_2 (U'_2 + v)$$

cancelling the terms involving v then gives

$$m_1 u'_1 + m_2 u'_2 = m_1 U'_1 + m_2 U'_2$$

In other words if momentum is conserved in the XYZ reference frame it is also conserved in the $X'Y'Z'$ frame. We note also the important point that the observation of the collision from the two frames has given us no information about their relative motion. Thus if we label the XYZ frame as the 'stationary' frame the observation of the collision has told us nothing about the velocity of the $X'Y'Z'$ frame.

A simple exercise (problem 1.3) shows that if we postulate that kinetic energy ($\frac{1}{2}mu^2$) is conserved in the collision when observed in the XYZ frame, then conservation of energy will also be found if the measurements are performed in the $X'Y'Z'$ frame. We may sum up our considerations in the following schematic form—if we denote the total momentum and energy as p and E respectively and the systems before and after the collision by subscripts i (initial) and f (final) respectively then

$$\begin{array}{cc} p_i = p_f & \xrightarrow[\text{transformation}]{\text{Galilean}} \quad p'_i = p'_f \\ E_i = E_f & \qquad\qquad\qquad E'_i = E'_f \end{array} \tag{1.1.5}$$

What is being said in these equations is that the laws of conservation of momentum and energy for a given physical process are independent of whether the reference frame of the observer is at rest or in a state of uniform motion. In general, reference frames in which free bodies do not experience any acceleration are called *inertial reference frames*.

We shall see later that the equations (1.1.5) have a more general validity than in Newtonian mechanics. We shall also find that equation (1.1.4) is true in relativistic mechanics provided that we reinterpret the role of the mass m. Indeed it is a noteworthy feature of Newtonian mechanics that no one suspected any flaw in their original formulation until Einstein developed the theory of special relativity, pointed out the error in treating m as a constant and then corrected this misconception (§4.2).

Finally we note that although we have so far only considered momenta and energies, forces will remain invariant in either frame. This follows from the relation

$$\text{force} = \text{mass} \times \text{acceleration}$$

and if we denote acceleration by a in the XYZ frame

$$a = \frac{du}{dt} = \frac{d^2x}{dt^2} \tag{1.1.6}$$

then in the $X'Y'Z'$ frame

$$a' = \frac{du'}{dt} = \frac{d}{dt}(u-v) = \frac{du}{dt} = a \tag{1.1.7}$$

since v is constant.

1.2   Revolutions in physics in the nineteenth century

The prime objective of the physicist is to explain the observed properties of the universe with the minimum possible number of physical laws and assumptions.

Until the beginning of the nineteenth century, electricity and magnetism were regarded as two independent subjects with certain similarities between them. The work of Oersted and Biot and Savart, in the early 1820s, which showed that electric currents produced magnetic fields, was later reinforced by that of Faraday which clearly demonstrated that changing magnetic fluxes could induce emfs. Thus the two subjects gradually merged into the single one—electromagnetism.

During the merger a curious feature emerged, namely that the ratio of the unit of charge in electrostatic units to that in electromagnetic was remarkably close to the value of the velocity of light. The law of Biot and Savart stated that the magnetic field B produced at a point P (figure 1.3) by a current of

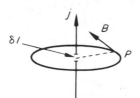

Figure 1.3

strength j flowing in an element of conductor of length δl is directly proportional to the product $j\,\delta l$ and inversely proportional to the square of the distance of separation, r, of P and δl

$$cB = j\frac{\delta l}{r^2} \tag{1.2.1}$$

and that the field acts at right angles to the plane containing j and r. The term c is a constant of proportionality in this expression. (Vector notation will be ignored at this point as it adds nothing to the argument.)

Now consider the implications of equation (1.2.1). We shall apply dimensional analysis to it; in addition to the conventional M, L and T for mass, length and time we shall use P and Q for magnetic pole strength and electric charge respectively with the classical definitions of Coulomb's law for magnetic and electrical forces

$$[\text{Force}] = \frac{[P^2]}{[L^2]} = \frac{[Q^2]}{[L^2]} \tag{1.2.2}$$

In equation (1.2.1) we then have

$$[j] = \frac{[Q]}{[T]} \qquad [B] = \frac{[P]}{[L^2]}$$

so that

$$[c] = \frac{[Q][L^2]}{[T][P][L]}\frac{1}{} = \frac{[Q][L]}{[P][T]} \tag{1.2.3}$$

and since Q and P have the same dimensions by virtue of (1.2.2) then c has the dimensions of a velocity.

The measurement of c was first carried out by Weber and Kohlrausch in 1856. They measured the charge on a Leyden jar. Firstly the potential (V) by an electrometer technique (electrostatic attraction) together with a calibration of its capacity (C) so that $Q = CV$ and secondly by determining the throw of a ballistic galvanometer (electromagnetic measure). The value obtained for c was

$$c = 3.1074 \times 10^8 \text{ m/s}$$

This result, so close to the known value for the velocity of propagation of light, was virtually the keystone in the bridge linking electromagnetism with yet another apparently independent branch of physics—the subject of optics. The first strong qualitative indication of a link had come in 1845 in the work of Faraday who showed that the plane of polarisation of a beam of polarised light could be rotated when the beam passed through a magnetised transparent substance. In Faraday's experiment he used a slab of glass between the poles of an electromagnet and found a rotation of the polarisation when the direction of the beam was parallel to that of the magnetic field.

The final linking of electromagnetism and optics in one uniform system was made by Maxwell in a series of papers between 1862 and 1865 culminating in a compact series of equations that have been written in Gaussian units, as these offer the greatest clarity in our present discussion

$$\text{curl } \mathbf{E} = -\frac{1}{c}\frac{\partial \mathbf{B}}{\partial t}$$

$$\text{curl } \mathbf{B} = \frac{1}{c}\frac{\partial \mathbf{E}}{\partial t} + 4\pi\mathbf{j} \qquad (1.2.4)$$

$$\text{div } \mathbf{E} = 4\pi\varrho$$

$$\text{div } \mathbf{B} = 0$$

where

$\mathbf{E}$ = electric field
$\mathbf{B}$ = magnetic field
ϱ = charge density
$\mathbf{j}$ = current density

These equations contain (admittedly, in a not very obvious way from mere inspection) all the known laws and observed behaviour in classical electromagnetism. Maxwell was also able to show that the constant c appearing in the equations represented the velocity of propagation of the electromagnetic waves. We have already seen that the experiment of Weber and Kohlrausch showed that c was numerically equal to the velocity of light. Thus light behaved like an electromagnetic radiation of high frequency. (The frequency is given by c/wavelength; the largeness of c and smallness of the wavelength implies that light possesses very high frequency.) Later work, notably that of Hertz (1883), provided experimental proof of the existence of electromagnetic waves and showed that they interfered like light waves and travelled with the velocity of light and thus confirmed the broad spectrum of Maxwell's work. Thus the three topics of electricity, magnetism and optics which were regarded as separate subjects at the beginning of the nineteenth century had been effectively merged into one.

1.3   Problems

Like all good theories, the work of Maxwell solved many problems but raised others. We have noted in §1.1 that the laws of Newtonian mechanics remain invariant under Galilean transformations; that is, they remain the same in form whether the observer is at rest relative to the process he is measuring or whether he is moving relative to it with a constant velocity. However, when the same transformation (1.1.2) was applied to Maxwell's equations they did not remain invariant.* Let us amplify this statement by considering a specific example—the equation of propagation of the electric field $\mathbf{E}$ in free space

$$\frac{\partial^2 \mathbf{E}}{\partial x^2} = \frac{1}{c^2}\frac{\partial^2 \mathbf{E}}{\partial t^2}$$

---

* The difficulty in fact was noted before it arose in Maxwell's equations. Maxwell quotes in his book (*A Treatise on Electricity and Magnetism*, vol. 2, p. 483, 3rd edition) the work of Gauss in 1835 (which was not published during his lifetime), in which Gauss concludes for Coulomb's law: 'two elements of electricity in a state of relative motion attract or repel one another, but not in the same way as if they are in a state of relative rest'.

A transformation must cause this equation to assume the form

$$\frac{\partial^2 \mathbf{E}'}{\partial x'^2} = \frac{1}{c^2} \frac{\partial^2 \mathbf{E}'}{\partial t'^2}$$

in order to satisfy the invariance condition. However, with some labour, one may show that the Galilean transformation (1.1.2) does not lead to the property

$$\frac{\partial^2}{\partial x^2} - \frac{1}{c^2} \frac{\partial^2}{\partial t^2} = \frac{\partial^2}{\partial x'^2} - \frac{1}{c^2} \frac{\partial^2}{\partial t'^2}$$

and so the invariance condition cannot be satisfied. We shall postpone the detailed discussion of the mathematical treatment of this problem until a later chapter (§6.2 and 6.3), where it will also be shown that the invariance condition is automatically satisfied in Einstein's theory of special relativity.

The difficulties with the laws of electrodynamics were compounded by two ideas which dominated the mode of thinking of physicists in the nineteenth century—the concept of an ether and the possibility of absolute reference frames. The latter concept went back at least to Newton who assumed the existence of an absolute space, which 'in its own nature, and without reference to anything external, remains always similar and immovable'*. The idea of an ether arose originally in the early work on optics, and was based on the concept that wave motion needed a medium (ether) in which to propagate and that this medium acted as a receptacle of energy in the passage of electromagnetic energy from one point to another. The ether was assumed to pervade all space. Maxwell's work was completely dominated by this central idea. Further it was believed that if the Galilean transformation was correct Maxwell's equations could only be valid in one reference frame and this was taken to be the reference frame of the ether. Thus the constant c in Maxwell's equations was believed to represent the *velocity of light relative to the ether*.

The theory of the ether reached its peak in the work of Lorentz[†] who developed a set of electromagnetic equations from a consideration of the interaction of electrons. The electrons were presumed to interact not directly with each other but with the ether in which they were embedded. The ether was assumed to be at rest in absolute space. Lorentz was able to show that the theory of electrons not only gave Maxwell's laws in the simplest situation, but that it also explained many phenomena in optics, including the rotation of the plane of polarisation of light in a magnetic field which we have mentioned in §1.2.

It should be noted that not all physicists were convinced of the existence or necessity of an ether, and Maxwell adopts a curiously defensive style at places in his *Treatise* when discussing the ether.

---

* *Principia*, p. 9.
[†] *Archs néérl.* **25**, 363, 1892.

Let us for the moment, however, assume the existence of an ether and a reference frame Σ which is stationary relative to it (figure 1.4). An observer

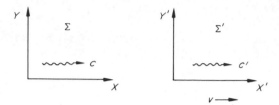

Figure 1.4

located in the ether frame will then measure the velocity of light as c. However if a second observer, located in frame Σ' which moves with velocity v along the X axis relative to Σ, also attempts to measure the velocity of the same light beam, then the Galilean transformation (1.1.3) tells us that a velocity

$$c' = c - v \qquad (1.3.1)$$

would be expected.

It is apparent that the relation given above allows us to determine v if we know c and c'. A number of experiments based on this principle were performed in the nineteenth century, but all failed to detect v. The most famous was that of Michelson in 1881[*] who exploited a suggestion made by Maxwell.[†] Maxwell pointed out that in principle one could determine the earth's velocity through the ether by determining the time difference which occurred when light travelled in opposite directions across the earth's surface.

Consider a system with three fixed points, A, B and C, in the earth's reference frame Σ', (figure 1.5). Let AC be at right angles to AB, and let there be a source of light at A and mirrors at B and C. We shall assume that the earth has a velocity v through the ether in the direction AB, and shall denote the ether reference frame as Σ.

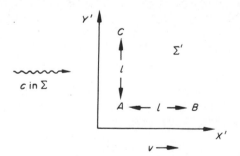

Figure 1.5

[*] *Am. J. Sci.*, **122**, 120, 1881. An improved version of this experiment was later performed in collaboration with E. W. Morley—*Am. J. Sci.*, **134**, 333, 1887.
[†] *Scientific papers*, vol. 2, p. 763, Dover, 1952.

According to the equation for the Galilean transformation (1.3.1) an observer located in Σ' will record two velocities for the light

$$c'_{A \to B} = c - v \quad c'_{B \to A} = c + v$$

where once again c is the velocity of light in the ether reference frame. The time for the passage of the light in the two directions is

$$t_{ABA} = \frac{l}{c - v} + \frac{l}{c + v} = \frac{2l}{c} \cdot \frac{1}{1 - v^2/c^2} \tag{1.3.2}$$

Now let us examine a second ray which travels from A to the mirror C and back to A. Let the time for the light to travel from A to C be t. During this time the mirror C will have moved a distance vt to the right, thus t will be determined by the Pythagoras theorem

$$c^2 t^2 = l^2 + v^2 t^2 \tag{1.3.3}$$

Therefore we find

$$t = \frac{l}{\sqrt{(c^2 - v^2)}} = \frac{l}{c} \cdot \frac{1}{\sqrt{(1 - v^2/c^2)}}$$

and since the same argument applies to the return journey, the total time is

$$t_{ACA} = \frac{2l}{c} \cdot \frac{1}{\sqrt{(1 - v^2/c^2)}} \tag{1.3.4}$$

Thus there is a time difference for the light to travel along the paths ABA and ACA

$$\delta t = t_{ABA} - t_{ACA}$$

The ingenious experiment of Michelson and Morley set out to determine v by measuring δt (the details of their experiment are given in appendix 1). They found no difference in time and suggested that $v = 0$.*

This result aroused considerable interest and controversy as it appeared to reject the concept of a stationary ether. One ingenious explanation in the context of an ether, was offered by Lorentz[†] who suggested that the diameter of the earth, and thus all other lengths, shortened in the direction v by an amount $\sqrt{(1 - v^2/c^2)}$, but remains unaltered at right angles to v. Thus the time t_{ABA} in figure 1.5 becomes

$$t_{ABA} = \frac{2l}{c} \frac{\sqrt{(1 - v^2/c^2)}}{1 - v^2/c^2} = \frac{2l}{c} \frac{1}{\sqrt{(1 - v^2/c^2)}} \tag{1.3.5}$$

---

* The quantitative results from this experiment, together with subsequent, more accurate, work are discussed in appendix 1.
† The same suggestion was made independently by Fitzgerald.

which is the same as t_{ACA} (1.3.4) and so no effect would be observed in the Michelson–Morley experiment.

Lorentz explained* his contraction factor in terms of a modification of molecular forces in a direction parallel to the motion of the earth through the ether just in the same manner as electromagnetic forces were assumed to change. Thus he concluded that not only the length l contracted in Michelson's apparatus but also the apparatus which measured the length l so that the effect was not detectable.

As we shall see later the Einstein theory of relativity also leads to contractions of length by a factor $\sqrt{(1-v^2/c^2)}$, but this result arises from a totally different set of assumptions.

1.4   The age of Einstein

The work of Lorentz and Fitzgerald offered a plausible explanation why attempts to detect the motion of the earth relative to the ether would fail. Lorentz went a stage further.[†] We mentioned at the beginning of §1.3 that Maxwell's laws of electrodynamics did not appear to be invariant under the Galilean transformation (1.1.2)

$$
\begin{aligned}
x' &= x - vt \\
y' &= y \\
z' &= z \\
t' &= t
\end{aligned}
$$

Thus experiments on, say, the interaction of two electrons could lead to different physical conclusions in two different reference frames.

Lorentz was able to show that if the transformations assumed the form

$$
\begin{aligned}
x' &= \frac{x - vt}{\sqrt{(1 - v^2/c^2)}} \\
y' &= y \\
z' &= z \\
t' &= \frac{t - vx/c^2}{\sqrt{(1 - v^2/c^2)}}
\end{aligned}
\tag{1.4.1}
$$

then Maxwell's equations in charge free space remained the same in all reference frames in uniform motion relative to each other. The equations (1.4.1) are now known as the equations of the *Lorentz transformation*, and physical equations and systems which remained invariant under this transformation are called *Lorentz invariants*. We shall give numerous examples in later chapters.

An inspection of equations (1.4.1) show that in situations where $v \ll c$ the transformation reduces to the Galilean one (1.1.2). Until the advent of atomic

---

* Lorentz's views (in translation) may be found in *The Principle of Relativity*, trans. W. Perrett and G. B. Jeffery, Methuen, 1923, p. 1.
† *Proc. Acad. Sci. Amst.* **6**, 809, 1904.

and nuclear physics no kinematic situations involved $v \to c$ and so there was no reason to suspect the incomplete nature of the Galilean transformation.

We shall see in the next chapter that equations of the Lorentz transformation form the basic core of Einstein's work on special relativity. However the physical models of the universe underlying the work of Lorentz and Einstein were completely opposed. Lorentz believed firmly in an absolute reference frame—the ether frame, and an absolute time scale despite the fact that he had got to the position where he had shown that motion through the ether was undetectable, since the measuring devices for length and time conspired to hide its presence, and that there was no need for a preferred reference frame. Einstein argued from the viewpoint that the concepts of an absolute space and time were all wrong. Once this assumption was made, the need for an ether in the Lorentz sense vanished.

One final major contributor to this period must be mentioned, namely Poincaré. In a paper published in 1904 * he cast doubt on the need for an ether and stated what is now described as the relativity principle — that the laws of physics should be the same in all reference frames which move in uniform motion with respect to each other. He also suggested that no velocity can exceed that of light and that an entirely new theory was necessary. However it was left to Einstein to make the final break with the previous notions of space and time and to provide that new theory.

PROBLEMS

1.1 A traveller paces up and down the aisle of a train at a speed of 2 miles per hour. If the train is moving along a straight track at 60 miles per hour, what is the traveller's speed with respect to the ground when he paces (a) in, (b) against, the direction of the train?

1.2 Two electrons are emitted in opposite directions from a piece of radioactive material at rest in the laboratory. If the speed of each electron is determined to be $0.8c$ (where c is the velocity of light) in the laboratory frame, what is the velocity of each electron in the rest frame of the other according to the Galilean transformation?

1.3 Verify the statement made in §1.1, in connection with the collision discussed in equation (1.1.4), namely that if the kinetic energy is conserved in the XYZ frame then it is also conserved in the $X'Y'Z'$ frame.

1.4 The scale diagram on page 13 depicts the collision of a ball a with another b, of equal mass, which is at rest. After the collision both balls move at 45° with respect to the direction of a. Draw how the collision appears in the reference frame of a'.

1.5 Verify that momentum and kinetic energy are conserved in the rest frames of b and a' in the previous problem.

1.6 Calculate δt in the Michelson–Morley experiment (§1.3). The equations (1.3.2) and (1.3.4) were obtained by assuming that the arm AB of the Michelson–Morley apparatus was parallel to the velocity v of the earth through the ether. Calculate δt if the arm AB is inclined at an angle θ. You may assume that $c \gg v$.

---

* *Bull. Sci. Math.* **28**, 302, 1904; *C.r. Acad. Sci.*, Paris **140**, 1504, 1905. The expression 'Lorentz transformation' first occurs in the latter paper.

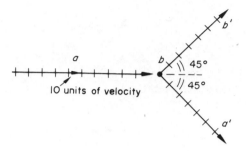

1.7 Verify that the Lorentz–Fitzgerald hypothesis (§1.3) still leads to a null result in the Michelson–Morley experiment if the arm AB is inclined at an angle θ to the velocity of the earth through the ether.

2

Einstein and relativity

2.1 Preamble

In 1905 Einstein published three remarkable papers—'Generation and transformation of light'* which proposes that light can be regarded as a stream of photons, 'Motion of suspended particles in the kinetic theory'† which is concerned with Brownian motion and 'Electrodynamics of moving bodies'‡ which we now regard as the foundation of the theory of special relativity.

Each paper is, in its own right, a major contribution to physics. It is a measure of Einstein's genius that his Nobel prize was awarded for the first paper. However, undoubtedly the most remarkable of the three is the 'Electrodynamics of moving bodies'. His paper of 1905 was submitted at the same time as that of Poincaré and had been written without his being aware of the paper of Lorentz in 1904. It contained not only the essential results of the

---

* *Annln Phys.* **17**, 132, 1905.
† *Ibid.* p. 549.
‡ *Ibid.* p. 891.

other two papers, but also demonstrated a totally different, and much more profound, understanding of the current problems in physics at that time.

2.2 The assumptions

The theory of special relativity is based on two assumptions. We shall quote them firstly in Einstein's own words.* At the beginning of his paper in 1905 he comments on the fact that in situations like the mutual interaction of a magnet and a conductor the observable phenomena depend on their relative motion and not the separate motion of each. He then continues:

'Examples of this sort, together with the unsuccessful attempts to discover any motion relative to the "light medium", suggest that the phenomena of electrodynamics as well as mechanics possess no properties corresponding to the idea of absolute rest. They suggest rather that...the same laws of electrodynamics and optics will be valid for all frames of reference for which the equations of mechanics hold good. We will raise this conjecture (the purport of which will hereafter be called the "Principle of Relativity") to the status of a postulate, and also introduce another postulate, which is only apparently irreconcilable with the former, namely, that light is always propagated in empty space with a definite velocity c which is independent of the state of motion of the emitting body. These two postulates suffice for the attainment of a simple and consistent theory of the electrodynamics of moving bodies based on Maxwell's theory for stationary bodies. The introduction of a "luminiferous ether" will prove to be superfluous inasmuch as the view to be developed will not require an "absolutely stationary space" provided with special properties...'.

Nowadays the physics of elementary particles has led us to the realisation of the existence of further basic interactions in addition to electrodynamics; so we generalise the first postulate whilst retaining the second in its original form. We shall therefore take as our basic assumptions:

(1) *The laws of physics are the same in all coordinate systems which move uniformly relative to one another.*

(2) *The velocity of light in empty space is the same in all reference frames and is independent of the motion of the emitting body.*

The second postulate has been subjected to a number of experimental tests, the most sensitive being that of Alvager, Farley Kjellman and Wallin [†] in 1964. These workers employed π^0 mesons as their source of light. The π^0 mesons are particles which decay to two photons (that is, light of very high frequency)

$$\pi^0 \rightarrow 2\gamma$$

In the experiment of Alvager and his co-workers the π^0 mesons were travelling with a velocity of $0.9975c$ and the velocity of the photons was measured along the direction of flight of the π^0 mesons. The result obtained was (2.9977

---

* These are available in translation in *The Principle of Relativity* trans. W. Perrett and G. B. Jeffery, p. 37, Methuen, 1923.

† *Phys. Lett.* **12**, 260, 1964.

$\pm 0.0004)$ 10^8 m/s which agrees well with the values for c obtained in experiments in which stationary sources were used.

2.3 The implications*

Consider a reference frame Σ (figure 2.1) with a light source located at O and a detector at D. We assume that the points O and D are stationary in Σ. Let a pulse of light be emitted from O; if the distance OD is x then the time for a light signal to travel from O to D will be

$$t = \frac{x}{c} \tag{2.3.1}$$

and this can be recorded by an observer stationed in Σ. Now let us assume that a second observer views this procedure from a reference frame Σ' which is moving parallel to the X axis. For convenience we shall assume that the

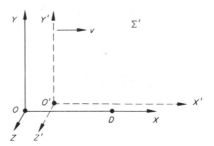

Figure 2.1

frames coincided at the time when the signal was emitted. The second observer measures the distance OD as x' and the principle of the constancy of the velocity of light implies that he will record the time as

$$t' = \frac{x'}{c} \tag{2.3.2}$$

Since Σ' is moving in the positive X direction x' is less than x and so t' must be less than t. Thus we can see that we can no longer accept that $t' = t$ as in the Galilean transformation (1.1.2).

Let us examine what the observers record in the frames Σ and Σ' in more detail. The observer stationed in Σ (in which the light source at O is at rest) sees the wave front of the light from O spreading as a sphere, and at any mo-

---

* Our discussion for the rest of this chapter will follow the spirit of a paper by Minkowski, available in translation in *The Principle of Relativity* (trans. W. Perrett and G. B. Jeffery), Methuen, 1923. Judging from Einstein's lectures (available in *The Meaning of Relativity*, A. Einstein (trans. E. P. Adams), Methuen, 1922), he also preferred Minkowski's treatment to his own.

ment in time, t, positions in the wave front are located by the relation

$$x^2 + y^2 + z^2 = c^2 t^2$$

or

$$x^2 + y^2 + z^2 - c^2 t^2 = 0 \tag{2.3.3}$$

The observer located in Σ' would also see the light spreading from O' as a sphere (recall that we postulated that O and O' coincided at the moment when the light pulse was emitted); thus the two postulates of Einstein lead to the following equation for the light sphere seen in Σ'

$$x'^2 + y'^2 + z'^2 - c^2 t'^2 = 0 \tag{2.3.4}$$

These two relations can only be satisfied if the two observers see *different* spheres. We illustrate this point schematically in figure 2.2, where the light spreads from sources O and O'.

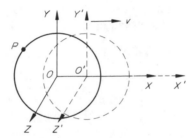

Figure 2.2

Now let us assume that Σ' moves with respect to Σ with a constant velocity v in the positive X direction. An inspection of figure 2.2 suggests that the components y, z will remain unaltered in going from the Σ to Σ' frames

$$y' = y \qquad z' = z$$

These equalities appear to be highly plausible from figure (2.2). However, the argument can be made more rigorously, but in doing so we must anticipate the discussion up to equation (2.3.9) in the text. The reader may ignore the following paragraph if he wishes and take the equality for granted.

Suppose y did change so that we had $y' = \phi(v) y$, where ϕ is some function of v. If space is homogeneous all directions in it must be equivalent, and so $\phi(v)$ must depend on the magnitude of v but not its direction, that is $\phi(v)$ must be an even function of v. Now supposing the observer in Σ reports on what the observer in Σ' is recording. The observer in Σ will say that Σ' is recording a length $y'' = \phi(-v) y'$. But this information is already recorded in Σ as y, consequently

$$y'' = y = \phi(-v) y' = \phi(-v) \phi(v) y$$

and so $\phi(-v)\phi(v)=1$. But $\phi(-v)=\phi(v)$ since we have already stated that $\phi(v)$ is a function of even powers of v; we therefore conclude $\phi(v)=1$.

We shall therefore concentrate on the relation

$$x^2 - c^2t^2 = x'^2 - c^2t'^2 = 0 \qquad (2.3.5)$$

which we obtain from (2.3.3) and (2.3.4). (Another possible relation is $(x^2 - c^2t^2) = \phi(x'^2 - c^2t'^2)$. However, it is easy to show that $\phi = 1$ by a similar argument to that which we applied to y.)

We now wish to relate x, x', t and t' (2.3.5). Let us assume that the relationship is linear (that is, that it involves first powers of terms only)

$$\begin{aligned} x' &= \gamma x + \delta t \\ x &= \varepsilon x' + \sigma t' \end{aligned} \qquad (2.3.6)$$

where γ, δ, ε and σ are constants independent of x, x', t and t'.

The transformations (2.3.6) must be linear to satisfy the condition that the relative velocity of Σ and Σ' is uniform (this condition is required by the first postulate of relativity—§ 2.2). Consider the first of equations (2.3.6); the position $x' = 0$ corresponds to the origin of the Σ' frame, hence

$$0 = \gamma x + \delta t \quad \text{or} \quad x = -\delta t/\gamma \qquad (2.3.7)$$

and the velocity of the Σ' frame is given by

$$v = \frac{\mathrm{d}x}{\mathrm{d}t} = -\frac{\delta}{\gamma} = \text{constant.} \qquad (2.3.8)$$

It is evident that any other power of t than one in (2.3.6) would destroy the requirement of constant v. Similarly any power of x other than one would lead to a non-uniform velocity which depended on position.

Next let us examine the Σ reference frame from the viewpoint of the Σ' frame. If an observer stationed in Σ states that the Σ' frame moves in the X direction with velocity $+v$, then one stationed in Σ' will say that the Σ frame moves with velocity $-v$ with respect to the X' axis. This statement is based on the assumption that there can be only one magnitude for the relative velocity and that all directions of space are equivalent. This principle is sometimes called the 'homogeneity of space'. The origin of the Σ frame is at the point $x = 0$, thus its motion can be represented in the Σ' frame as

$$0 = \varepsilon x' + \sigma t'$$

from the second of equations (2.3.6), and so

$$\frac{\mathrm{d}x'}{\mathrm{d}t'} = -\frac{\sigma}{\varepsilon} = -v \qquad (2.3.9)$$

from our arguments above.

Let us now try to find solutions for γ, δ, ε, σ. The equations (2.3.8) and (2.3.9) imply that we have transformed equations (2.3.6) into

$$x' = \gamma (x - vt)$$
$$x = \varepsilon (x' + vt') \tag{2.3.10}$$

We may eliminate x' from these equations by substitution and solve for t'

$$t' = \gamma \left[t - \frac{x}{v} \left(1 - \frac{1}{\gamma \varepsilon} \right) \right]$$

then if we substitute for x' and t' in the right member of (2.3.5) we find

$$x^2 \left[\gamma^2 - \frac{c^2 \gamma^2}{v^2} \left(1 - \frac{1}{\gamma \varepsilon} \right)^2 \right] + xt \left[\frac{2c^2 \gamma^2}{v} \left(1 - \frac{1}{\gamma \varepsilon} \right) - 2\gamma^2 v \right] - t^2 \left[c^2 \gamma^2 - v^2 \gamma^2 \right] = 0 \tag{2.3.11}$$

We can now compare this result with the left side of (2.3.5) and equate coefficients. We then have

for $\qquad\qquad\qquad x^2 \qquad\qquad \gamma^2 - \frac{c^2 \gamma^2}{v^2} \left(1 - \frac{1}{\varepsilon \gamma} \right)^2 = 1$

$$xt \qquad\qquad \frac{2c^2 \gamma^2}{v} \left(1 - \frac{1}{\varepsilon \gamma} \right) - 2\gamma^2 v = 0$$

$$t^2 \qquad\qquad c^2 \gamma^2 - v^2 \gamma^2 = c^2.$$

Considering firstly the coefficient of t^2 we find

$$\gamma = \frac{1}{\sqrt{(1 - v^2/c^2)}} \tag{2.3.12}$$

(we choose the positive solution so that the equation (2.3.10)

$$x' = \gamma (x - vt)$$

makes sense in the limit $v/c \to 0$). Next consider the coefficient of xt

$$\frac{2c^2 \gamma^2}{v} \left(1 - \frac{1}{\gamma \varepsilon} \right) - 2\gamma^2 v = 0$$

which leads to

$$\frac{1}{\gamma \varepsilon} = 1 - \frac{v^2}{c^2}$$

and so from equation (2.3.12)

$$\varepsilon = \gamma = \frac{1}{\sqrt{(1 - v^2/c^2)}} \tag{2.3.13}$$

The proof that the coefficient of x^2 is one in equation (2.3.11) is left as an exercise for the reader.

The substitution our results for ε and γ into equation (2.3.10) then gives us

$$x' = \gamma(x - vt) = \frac{x - vt}{\sqrt{(1 - v^2/c^2)}}$$
$$x = \gamma(x' + vt') = \frac{x' + vt'}{\sqrt{(1 - v^2/c^2)}} \qquad (2.3.14)$$

and a comparison with equations (1.4.1) shows that we have obtained the first of the equations of the Lorentz transformation. It is then a simple matter to eliminate x or x' from these equations and get expressions for t and t'. The full Lorentz transformation equations then become

$$
\begin{array}{ll}
x' = \gamma(x - vt) & x = \gamma(x' + vt') \\[2mm]
y' = y & y = y' \\[2mm]
z' = z & z = z' \\[2mm]
t' = \gamma(t - vx/c^2) & t = \gamma(t' + vx'/c^2)
\end{array}
\qquad (2.3.15)
$$

where

$$\gamma = \frac{1}{\sqrt{(1 - v^2/c^2)}}$$

and v is the velocity of the Σ' frame when measured from Σ.

2.4 Intervals

In what follows we shall often have to refer to an *event*. This is some occurrence whose position in space and time is simultaneously measured.

Position in space requires the stipulation of three coordinates, and time one value. In relativity theory the introduction of a fourth coordinate with the same dimensions as the spatial coordinates is found to be extremely convenient, and this is done by introducing the parameter ct. The event can then be registered as a point in a fictitious (but mathematically convenient) four-dimensional space. The point is often called a *world point*.

Now consider two events, which we label a and b, and let their world points be located at x_a, y_a, z_a, ct_a and x_b, y_b, z_b, ct_b respectively. The *interval* S_{ab} between the events is defined to be

$$S_{ab}^2 = c^2(t_a - t_b)^2 - (x_a - x_b)^2 - (y_a - y_b)^2 - (z_a - z_b)^2. \qquad (2.4.1)$$

One example of an interval can be seen in figure 2.2; the points O and P represent respectively the source and detector of a light pulse. If we say the world point associated with O has coordinates $x_b=y_b=z_b=ct_b=0$ then it is apparent from equation (2.3.3)

$$x^2+y^2+z^2-c^2t^2=0$$

that the interval $S_{ab}=0$ in this case and furthermore the same condition applies at all other points on the light sphere.

In addition to the system displayed in figure 2.2 we can examine the intervals between events which are not connected by a light signal. An inspection of equation (2.4.1) shows that these divide into two parts

$$\begin{aligned}S_{ab}^2>0 \text{ that is } & \quad c^2(t_a-t_b)^2>(x_a-x_b)^2+(y_a-y_b)^2+(z_a-z_b)^2\\ S_{ab}^2<0 \text{ that is } & \quad c^2(t_a-t_b)^2<(x_a-x_b)^2+(y_a-y_b)^2+(z_a-z_b)^2\end{aligned} \quad (2.4.2)$$

In terms of figure 2.2 pairs of events which have $S_{ab}^2>0$ lie inside the light sphere (if we locate a or b at the origin O—which one we choose is merely a matter of labelling and so b will be selected); on the other hand $S_{ab}^2<0$ lies outside the light sphere. Intervals with $S_{ab}^2>0$ and $S_{ab}^2<0$ are said to be *time-like* and *space-like* respectively

$$\begin{aligned}S_{ab}^2>0 & \quad \text{time-like intervals}\\ S_{ab}^2<0 & \quad \text{space-like intervals.}\end{aligned} \quad (2.4.3)$$

2.5 The light cone

We now take the process a little further. We shall again assume that the coordinates of the world point b are $x_b=y_b=z_b=ct_b=0$, that is, the origin O of the reference frame, so that

$$S_{ab}^2=c^2t_a^2-x_a^2-y_a^2-z_a^2$$

and also introduce the symbol

$$r_a^2=y_a^2+z_a^2$$

and axis R (figure 2.3). The reason for the introduction of r is so that we can present a visual conception of a four-dimensional system, therefore we have

$$S_{ab}^2=c^2t_a^2-x_a^2-r_a^2 \quad (2.5.1)$$

Now if $S_{ab}^2=0$, then

$$c^2t_a^2=x_a^2+r_a^2 \quad (2.5.2)$$

and a plot of values of ct_a, x_a and r_a satisfying this condition produces two cones as in figure 2.3 (here T refers to the time axis). It is evident that the sides of the cones are at $45°$ to the cT axis. The uppermost cone is associated with positive values of ct_a, that is, $ct_a>0$, whilst the lower cone involves negative ct_a, that is, $ct_a<0$. Let us now examine their physical significance. We located

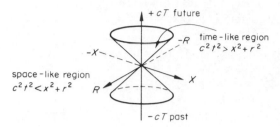

Figure 2.3

our event b at the origin O with coordinates $x_b = r_b = ct_b = 0$, thus if event a is later in time than b we have $ct_a > 0$; this means that event a is in the *future* relative to b. Conversely if event a is earlier in time than b, and therefore is in the *past* relative to b, then $ct_a < 0$. Thus the upper cone contains events which are in the future relative to O, whilst the lower contains events that are in the past. We must bear in mind, however, that so far we have been talking about events which lie on the surface of the cone and are connected by distances which correspond to the product of the time interval and the velocity of light (2.5.2). Such events are said to be *on the light cone*. However an inspection of equation (2.5.2) in conjunction with (2.4.2) and (2.4.3) shows that events which lie inside the cone, namely

$$c^2 t_a^2 > x_a^2 + r_a^2 \tag{2.5.3}$$

are separated from the origin by time-like intervals ($S_{ab}^2 > 0$), whilst space-like intervals will lie outside the cone

$$c^2 t_a^2 < x_a^2 + r_a^2. \tag{2.5.4}$$

If the interval between two events is time-like (that is, a lies inside the light cone with respect to b in figure 2.3), then they can be causally related to each other. Causally related events are of a type where an event at point b can cause a second event at point a. For example, a command signal can be sent from a transmitter located at b (figure 2.4) to a receiver at a along an electric

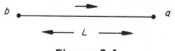

Figure 2.4

cable of length L*. An event can then occur at a as a result of the signal. The velocity, u, of transmission of signals along electric cables is less than that of light in free space because of the physical characteristics of the cable. Thus the time for the transmission of the signal L/u is greater than L/c, where c is as usual the velocity of light in free space. The command signal could also be

---

* In terms of components along the X, Y, Z axes
$$L^2 = x_a^2 + r_a^2 = x_a^2 + y_a^2 + z_a^2.$$

passed between a and b in a time L/c by dispensing with the cable and using a radio or light pulse. However it is not possible to pass a signal in a time less than L/c. This is because velocities greater than that of light in free space are not physically possible. The reason for this can be seen in the form of the equations for the Lorentz transformation (2.3.15). It is evident that if $v>c$ in these equations then

$$\gamma = \frac{1}{\sqrt{(1-v^2/c^2)}}$$

becomes imaginary and we pass into a world with imaginary lengths and times. The question of maximum possible velocities will be raised again in §3.7.

Thus if a signal is sent from b with a velocity c an event at a with coordinates x_a, r_a, ct_a cannot be causally related to it if $L>ct_a$, and since $L^2=x_a^2+r_a^2$ events with no causal relation must lie outside the light cone (compare our relation for space-like intervals (2.5.4)).

2.6   Invariance properties of the interval

We should now like to approach the Lorentz transformation again from a different angle. Let us go back once again to our definition of the interval (2.4.1)

$$S_{ab}^2 = c^2 (t_a - t_b)^2 - (x_a - x_b)^2 - (y_a - y_b)^2 - (z_a - z_b)^2 .$$

We now put our notions of space and time on a more equal footing by introducing the definitions

$$
\begin{aligned}
x_a - x_b &= x_1 \\
y_a - y_b &= x_2 \\
z_a - z_b &= x_3 \\
ict_a - ict_b &= x_4 \quad i=\sqrt{-1}
\end{aligned}
\tag{2.6.1}
$$

The idea of introducing an imaginary fourth component was originally due to Poincaré.* Thus our definition of the interval now becomes

$$-S_{ab}^2 = x_1^2 + x_2^2 + x_3^2 + x_4^2 = \sum_{\lambda=1}^{4} x_\lambda^2 \tag{2.6.2}$$

We now show that the above equation has interesting geometrical possibilities. Firstly consider the equations which define a circle and sphere. (r and R are the radii of circle and sphere respectively.)

$$
\begin{aligned}
r^2 &= x^2 + y^2 &&\equiv \sum_{\lambda=1}^{2} x_\lambda^2 \\
R^2 &= x^2 + y^2 + z^2 &&\equiv \sum_{\lambda=1}^{3} x_\lambda^2
\end{aligned}
\tag{2.6.3}
$$

---

* *C.r. Acad. Sci. Paris* **140**, 1504, 1905.

An inspection of the above equations and comparison with (2.6.2) shows that the latter defines a four-dimensional sphere of radius iS_{ab}. The presence of $i = \sqrt{-1}$ should not put the reader off; recall that S_{ab}^2 can be both positive and negative according to our discussion of §2.4 and so for space-like intervals iS_{ab} is actually a real number.

We now make use of an important property of circular geometry, namely that if the axes are rotated the radii remain unaltered. We illustrate this point with the circle in figure 2.5(a)—it can be seen that the rotation of axes through an angle θ has left the length of OP unchanged.

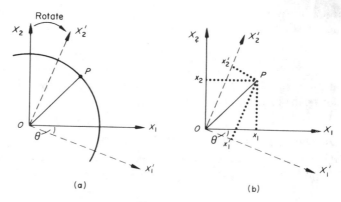

(a) (b)

Figure 2.5

Now consider the effect of the rotation on the components of OP; the point P has coordinates $x_1 (\equiv x)$ and $x_2 (\equiv y)$ in figure 2.5(b) so that

$$OP^2 = x_1^2 + x_2^2 \qquad (2.6.4)$$

Now elementary trigonometry yields the equations

$$x_1 = (x_1'/\cos\theta) + x_2 \tan\theta \qquad (2.6.5)$$
$$x_2' = (x_2/\cos\theta) + x_1' \tan\theta$$

and so we can rewrite the first equation as

$$x_1' = x_1 \cos\theta - x_2 \sin\theta$$

Hence the second equation in (2.6.5) becomes

$$x_2' = (x_2/\cos\theta) + (x_1 \cos\theta - x_2 \sin\theta) \tan\theta$$
$$= x_1 \sin\theta + x_2 (1 - \sin^2\theta)/\cos\theta$$
$$= x_1 \sin\theta + x_2 \cos\theta$$

Thus we have the two equations

$$x_1' = x_1 \cos\theta - x_2 \sin\theta$$
$$x_2' = x_1 \sin\theta + x_2 \cos\theta \qquad (2.6.6)$$

It is obvious from figure 2.5 that the length OP has remained unaltered by the rotation of axes. However we shall provide a check on equations (2.6.6) by verifying this fact algebraically

$$
\begin{aligned}
OP^2 &= x_1'^2 + x_2'^2 \\
&= (x_1 \cos\theta - x_2 \sin\theta)^2 + (x_1 \sin\theta + x_2 \cos\theta)^2 \\
&= x_1^2 (\cos^2\theta + \sin^2\theta) + x_2^2 (\sin^2\theta + \cos^2\theta) \\
&= x_1^2 + x_2^2
\end{aligned}
\tag{2.6.7}
$$

We also note in passing that equivalent formulae to (2.6.6), for the rotation of axes, could also have been obtained by rotating OP in an anticlockwise direction through an angle θ.

If we examine the equation of a sphere (2.6.3)

$$
R^2 = x^2 + y^2 + z^2 = \sum_{\lambda=1}^{3} x_\lambda^2
$$

it is evident that rotations in the XY, YZ, ZX, or equivalently the $X_1 X_2$, $X_2 X_3$, $X_3 X_1$, planes, will leave R^2 unaltered. This is easily mentally visualised, or, for example, with the aid of equation (2.6.7) we can write for a rotation in the $X_1 X_2$ plane

$$
\begin{aligned}
R^2 &= x_1^2 + x_2^2 + x_3^2 \\
&= x_1'^2 + x_2'^2 + x_3^2
\end{aligned}
\tag{2.6.8}
$$

and in general for more than one rotation

$$
\begin{aligned}
R^2 &= x_1'^2 + x_2'^2 + x_3'^2 \\
&= x_1^2 + x_2^2 + x_3^2
\end{aligned}
\tag{2.6.9}
$$

Now let us consider the four-dimensional sphere. It is left as an exercise for the reader to show that rotation in six planes is possible. The four-dimensional equivalent of equation (2.6.9) for multiple rotations is (compare (2.6.2))

$$
-S_{ab}^2 = x_1'^2 + x_2'^2 + x_3'^2 + x_4'^2 = \sum_{\lambda=1}^{4} x_\lambda'^2 = \sum_{\lambda=1}^{4} x_\lambda^2
\tag{2.6.10}
$$

We shall now concentrate on the implications of the above relation for a rotation in the $X_1 X_4$ plane (that is, in the XT plane). The components x_2 and x_3 will remain unaltered and equations (2.6.6) can easily be adapted to the present situation without further geometry

$$
\begin{aligned}
x_1' &= x_1 \cos\theta - x_4 \sin\theta \\
x_4' &= x_1 \sin\theta + x_4 \cos\theta
\end{aligned}
\tag{2.6.11}
$$

Now let us link the coordinates x with the two reference frames, Σ and Σ' which we have used previously. Once again we make the assumption that Σ' is moving with a uniform velocity v along the X_1 axis of Σ. We recall from equation (2.6.1) that x represents the 'distance' between two world points a

and b as shown in figure 2.6(a).

Let the point b represent the position in space and time where the origins of the Σ and Σ' frames coincided; and let the point a represent the spatial origin of the Σ' frame at subsequent times. Thus if the Σ' frame moves with

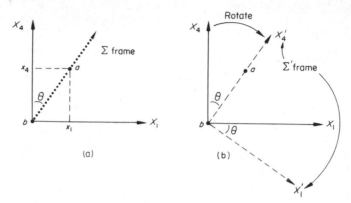

Figure 2.6

constant velocity v with respect to the X_1 axis of the Σ frame, then as a function of time in the Σ frame the point a traces out the dotted line illustrated in figure 2.6(a), and

$$\frac{x_1}{x_4}=\tan\theta=\frac{x_a-x_b}{ic(t_a-t_b)}=\frac{x_a}{ict_a}=\frac{vt_a}{ict_a}=\frac{-iv}{c}$$

Now consider a rotation of axes through the angle θ so that a lies on the x_4' axis at all times (figure 2.6 (b)). We see immediately that $x_1'=0$ at all times, and so the first line in equations (2.6.11) yields

$$x_1'=0=x_1\cos\theta-x_4\sin\theta \quad \text{or} \quad x_1/x_4=\tan\theta$$

Thus we find the relative velocity of the two reference frames is related to the tangent of the angle of rotation of the space-time plane.

Thus if we write the ratio v/c as β

$$\tan\theta=-i\frac{v}{c}=-i\beta \tag{2.6.12}$$

then the properties of the Pythagoras triangle (figure 2.7) give

$$\cos\theta=\frac{1}{\sqrt{(1-\beta^2)}}=\gamma \qquad \sin\theta=\frac{-i\beta}{\sqrt{(1-\beta^2)}}=-i\beta\gamma \tag{2.6.13}$$

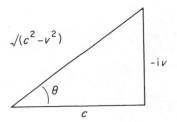

Figure 2.7

and substitution into equations (2.6.11) again yields the Lorentz transformation relations (2.3.15)

$$
\begin{array}{lll}
x'_1 = \gamma x_1 + i\beta\gamma x_4 & \equiv & x' = \gamma(x - vt) \\
x'_2 = x_2 & \equiv & y' = y \\
x'_3 = x_3 & \equiv & z' = z \\
x'_4 = -i\beta\gamma x_1 + \gamma x_4 & \equiv & t' = \gamma(t - vx/c^2)
\end{array}
\tag{2.6.14}
$$

and similarly

$$
\begin{array}{l}
x_1 = \gamma x'_1 - i\beta\gamma x_4 \\
x_2 = x'_2 \\
x_3 = x'_3 \\
x_4 = i\beta\gamma x'_1 + \gamma x'_4
\end{array}
\tag{2.6.15}
$$

where we have used the fact that x_2 and x_3 remain invariant under rotations in the $X_1 X_4$ plane. The reader should not bother about the fact that we are using complex numbers in (2.6.14) and (2.6.15)—they are merely of mathematical convenience and have no effect on the physics arguments as the $\sqrt{-1} = i$ terms always disappear in real physical situations.

Thus the invariance of the interval under rotations in space-time implies that its components must transform in the manner of the Lorentz transformation. We leave the cross check, that the substitution of equations (2.6.14) into (2.6.10) satisfies the relation

$$
-S_{ab}^2 = \sum_{\lambda=1}^{4} x'^2_\lambda = \sum_{\lambda=1}^{4} x^2_\lambda
$$

as an exercise for the reader. We shall return to the important principle of invariance repeatedly in later chapters.

PROBLEMS

2.1 Show that the wave front from a pulse of light cannot be a sphere in two reference frames which move with relative velocity v (figure 2.2 and related equations), if the reference frames are related by the Galilean transformation (1.1.2). Make the cross check that the Lorentz transformation (2.3.15) satisfies this condition.

2.2 Check equations (2.3.11) and (2.6.5).

2.3 In order to gain familiarity with the factor $\gamma = 1/\sqrt{(1-\beta^2)}$, which occurs repeatedly in Lorentz transformations, calculate γ for $\beta = 0, 0.2, 0.4, 0.6, 0.8, 0.9, 0.99$.

2.4 A reference frame Σ' passes a frame Σ with a velocity of $0.6c$ in the X direction. Clocks are adjusted in the two frames so that $t = t' = 0$ at $x = x' = 0$.
   (a) An event occurs in Σ with space–time coordinates $x_1 = 50$ m, $t_1 = 2 \times 10^{-7}$ s. What are its coordinates in Σ'?
   (b) If a second event occurs at $x_2 = 10$ m, $t_2 = 3 \times 10^{-7}$ s in Σ what is the difference in time between the events as measured in Σ'?

2.5 A spaceship of length 100 m in its own rest frame Σ' passes a second spaceship Σ at a relative speed of $\sqrt{(3)}\, c/2$ and on a parallel course. When the centre of Σ' passes an observer located in Σ, a crew member of Σ' simultaneously fires very short bursts from two lasers mounted perpendicularly at the ends of Σ'. Assuming that the laser beams travel negligibly short distances, calculate the positions at which marks appear on the hull of Σ. You may use the event of the two observers being adjacent as the reference points $x = x' = 0$, $t = t' = 0$, and can assume that Σ is of sufficient length that the laser beams will strike its hull.

2.6 Two spaceships, each of length l as measured in their own reference frames pass each other on a parallel course. If an observer located at the front of one of the ships measures a time interval T for the other ship to pass him show that the relative velocity v of the two ships can be determined from the formula

$$\frac{l}{v\sqrt{(1-v^2/c^2)}} = T$$

2.7 A reference frame Σ' moves with a velocity v in the X direction relative to a frame Σ (assume that the axes of the two systems are parallel). A rigid rod lies in the XY plane and moves with a velocity u, with components u_x and u_y, with respect to Σ. If the instantaneous projections of its length in Σ and Σ' are denoted by l_x, l_y and l'_x, l'_y respectively, show that

$$l_x = \gamma l'_x (1 - v u_x/c^2)$$
$$l_y = l'_y - \gamma v u_y l'_x/c^2$$

2.8 Consider the two events $x_a = y_a = z_a = 0$, $t_a = 0$ and $x_b = 6c$, $y_b = z_b = 0$, $t_b = 8$ s. Is the interval between the two events space-like or time-like?

2.9 Two events are recorded with coordinates

$$x_a = X, \; y_a = z_a = t_a = 0$$
$$x_b = y_b = z_b = t_b = 0$$

in an inertial frame Σ. Show that the interval between the two events is space-like, and that by a suitable choice of reference frame their spatial separation may be made to vary between X and infinity.

3

Applications of the Lorentz transformation

In this chapter we shall examine some of the implications of the Lorentz transformation. The effects we shall discuss are completely outside normal everyday experience, since they only start to manifest themselves at velocities close to that of light (even an astronaut orbiting the earth is only moving with a speed $\sim 2 \times 10^{-5} c$). Nevertheless the effects of the Lorentz transformation are part of the day to day experience for one small group in society—the high-energy physicists. We shall therefore start this chapter with a problem which occurs frequently in high-energy physics—the birth and death of a meson.

3.1 The life span of a meson

The bombardment of nuclear targets by very high-energy protons, electrons or photons from the big particle accelerators has revealed the existence of many types of meson. For our present purposes we shall describe a meson as a system which decays radioactively, sometimes through a chain of mesons, and the final stable decay products are either electrons, neutrinos or photons (or a combination of them). For example the pion (a meson with mass ~ 270 times that of the electron) normally decays as

$$\pi \to \mu \nu$$
$$\downarrow$$
$$e \nu \bar{\nu}$$

where the symbol μ denotes the muon (a particle with radically different nuclear properties to that of the pion and mass ~ 208 times that of the electron), e the electron and ν, $\bar{\nu}$ are neutrinos.

The mesons satisfy the normal law of radioactive decay, which states that the number of particles remaining after a time t is

$$N(t) = N\,e^{-t/\tau} \tag{3.1.1}$$

where N is the number of mesons present at time $t = 0$ and τ is the mean lifetime of the mesons.

Now consider the birth and death of a meson. We shall assume that the meson travels across a laboratory and shall examine its life span in the reference frames of the laboratory and the meson—we shall denote coordinates in the latter frame by primed symbols. In the meson reference frame an observer records the birth at the world point b at time t'_b and its death at world point a at a time t'_a. The point a and b have the same spatial position for an observer stationed in the meson reference frame (that is, $x'_{1a} = x'_{1b} = 0$,

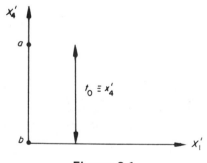

Figure 3.1

as in figure 3.1; the same situation also applies to the X'_2 and X'_3 axes). The time span

$$t'_a - t'_b = t_0 \tag{3.1.2}$$

measured for a system at rest in any reference frame is called the *proper time*. (We use the convention that measurements performed in a system's rest frame are designated by the subscript 0.)

In the laboratory frame an observer records the two events a and b at different spatial points with a time span

$$t_a - t_b = t$$

If the meson moves with a uniform velocity v in the laboratory frame, then the spatial separation of a and b is given by

$$(x_a - x_b)^2 + (y_a - y_b)^2 + (z_a - z_b)^2 = v^2 (t_a - t_b)^2 = v^2 t^2$$

The requirement of the invariance of the interval between events in different reference frames (2.6.10) then gives

$$\begin{aligned} -S_{ab}^2 &= -c^2 (t_a' - t_b')^2 = -c^2 t_0^2 \\ &= (x_a - x_b)^2 + (y_a - y_b)^2 + (z_a - z_b)^2 - c^2 (t_a - t_b)^2 \\ &= (v^2 - c^2)\, t^2 \end{aligned} \qquad (3.1.3)$$

hence

$$\boxed{\; t = \frac{t_0}{\sqrt{(1 - v^2/c^2)}} = \gamma t_0 \;} \qquad (3.1.4)$$

An inspection of (3.1.3) shows that there is an ambiguity of plus or minus signs when taking the square roots. The reason for the choice of the plus sign should become obvious after a discussion of causality in §3.4.

Equation (3.1.4) is the *time dilatation relation* of the theory of special relativity. An inspection of the above relation, together with the assumption that the laws of physics are valid in all reference frames, implies that the same equation must hold between the mean lives in equation (3.1.1)

$$\tau = \gamma \tau_0 \qquad (3.1.5)$$

A check on this equation has been made both for pions* and mesons.†
In both cases the time distribution (3.1.1) was measured (1) in the laboratory frame with particles in flight (t) and (2) with particles which had been brought to rest before decay so that the reference frames of the laboratory and particle coincided (t_0). The parameter γ was measured by determining the momentum of the particles. We anticipate §4.2 by stating that the momentum p of a particle of rest mass m_0 is given by

$$p = \gamma m_0 v$$

where v is the velocity of the particle. In the case of the pions γ was 2.4 and

---

* GREENBERG, A. J., AYRES, D. S., CORMACK, A. M., KENNEY, R. W., CALDWELL, D. O., ELINGS, V. B., HESSE, W. P., and MORRISON, R. J., *Phys. Rev. Lett.* **23**, 1267, 1969.
† BAILEY, J., and PICASSO, E., *Prog. in nucl. Phys.*, **12**, 43, 1970.

for the muons 12. The relation (3.1.5)

$$\tau = \gamma \tau_0$$

was verified to accuracies of 0.4 and 1.2 per cent for pions and muons respectively.

Next consider how each observer measures distances. The observer stationed in the meson reference frame states that the birth and death of the meson took place at the same spatial point in that frame. However, he can also describe the process in relation to the meson's position in the laboratory frame. He sees the laboratory frame moving past him with a velocity v and thus asserts that meson travels a distance

$$l' = v t_0$$

in the laboratory. Now the observer stationed in the laboratory also agrees that the two frames moved past each other with relative velocity v (this is an essential feature of the theory of relativity—both observers agree about the magnitude of the relative velocity, only its sign changes according to one's viewpoint (§2.3)). Thus he states that the meson moves a distance*

$$l_0 = v t$$

and so from (3.1.4)

$$v = \frac{l'}{t_0} = \frac{l_0}{t} = \frac{l_0}{\gamma t_0} \tag{3.1.6}$$

hence

$$\boxed{l' = \frac{l_0}{\gamma} = l_0 \sqrt{(1 - v^2/c^2)}} \tag{3.1.7}$$

This is the *Lorentz contraction* relation of special relativity—distances which are measured from a moving frame appear to be shorter than when the same distances are measured in a stationary frame.

Whilst the meson carries its own clock—its lifetime, it has no convenient devices for measuring length attached to it and so no direct experimental verification of the Lorentz contraction relation has proved to be possible.

There is, however, an interesting corollary attached to the relation for the Lorentz contraction. Consider the number of radioactive particles remaining

---

* At first sight the subscript 0 may appear to be in the wrong equation. We shall use the convention, however, that subscripts 0 appear when a measurement is made in a system's rest frame. Thus the length l_0 appearing in (3.1.6) is one which has been determined in the rest frame of the laboratory, whilst l' represents the distance an observer (travelling with the meson) asserts that he covered in the laboratory system.

after a time t_0 in their own rest frame (compare (3.1.1)).

$$N(t_0) = N \exp(-t_0/\tau_0)$$

but by equation (3.1.6)

$$t_0 = \frac{l_0}{\gamma v}$$

consequently in a reference frame in which the particles are moving with velocity v

$$N(l_0) = N \exp(-l_0/\gamma v\tau_0) \tag{3.1.8}$$

Thus in addition to living for a mean lifetime $\gamma\tau_0$ in the laboratory frame the particles travel for a mean distance $\gamma v\tau_0$ in it (as opposed to $v\tau_0$ when the same distance is examined from the particle's own rest frame).

One consequence of this relationship is the occurrence of muons at sea level in the cosmic radiation. The primary cosmic radiation, entering from outer space quickly disappears in nuclear collisions with the air atoms at the top of the earth's atmosphere. Pions are produced in these collisions and quickly decay to muons. The muons have a mean life of $2 \cdot 20 \times 10^{-6}$ s in their own reference frame and so even if they travelled with the velocity of light the mean distance covered would be $\sim 3 \times 10^8 \times 2 \times 10^{-6}$ m $= 600$ m. The dilatation factor γ ensures that a substantial fraction of them cover the ~ 20 km distance between the top of the atmosphere and sea level.

The dilation principle is also exploited in producing intense beams of high-energy mesons from the big particle accelerators. In the absence of the dilatation effects these mesons would decay close to their point of production; because of it the meson beams travel relatively long distances and can be engineered to suit specific experiments.

3.2   Clocks and rods

It is worth pausing to consider the implications of what we mean when we talk about time dilatation and Lorentz contraction. Distances are measured in terms of an internationally defined unit—the metre. Let us suppose that the distance l_0 in equation (3.1.7) was N metres and was represented by N rods, each of length 1 metre, placed end to end. The observer in the moving reference frame would claim that he had also passed N rods, but each of length $\sqrt{(1 - v^2/c^2)}$ metres.

Similarly, if the observer in the moving reference frame was equipped with a rod 1 metre long (defined at some time when both reference frames were at rest with respect to each other), then the observer in the rest frame would measure the rod to be $\sqrt{(1 - v^2/c^2)}$ metres. This statement can easily be proved explicitly, but it follows immediately from the principle of relativity, since which frame is defined to be at rest and which in motion is a matter of definition for the observers situated in each frame.

Thus it is the scale of length which changes in measuring distance in one moving reference frame from another.

Similarly, in measuring time, both observers would claim they measured the life span of the meson for n units of time, but these units change in going from one reference frame to the other. The form of clock used in making the measurement is irrelevant (provided that it does not rely on possible external reference frames like the simple pendulum). If different types of clock behaved in different ways in response to uniform motion, we should have a method of detecting that motion without reference to external frames—this would be at variance with the principle of relativity. Thus our meson clock is as good as any other in establishing the phenomenon of time dilation.

3.3   Time dilatation and Lorentz contraction

In this section we shall re-examine the conclusions of §3.1 from a different viewpoint by using the equations of the Lorentz transformation (2.3.15). Once again our world points b and a will be the birth and death points respectively of the meson as an aid to thought pictures.

We have seen in §2.6 that the equations of the Lorentz transformation can be generated by a rotation of space-time axes. This operation is performed

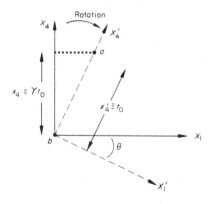

Figure 3.2

in figure 3.2. Recalling, from (2.6.13), that $\cos\theta = \gamma$ in this figure we see immediately that

$$x_4 = x_4' \cos\theta = \gamma x_4' \quad \text{hence} \quad t = \gamma t_0 \tag{3.3.1}$$

since

$$x_4 = ict \quad x_4' = ict_0$$

It should be noted that the distance x_4 is shorter than x_4' in figure 3.2. This is apparently in contradiction to the notion of time dilation. However we must remember that $\cos\theta$ is a hyperbolic cosine and hence greater than one since $v/c < 1$.

Instead of considering rotations we could have obtained the formula for the time dilatation by writing the equations of the Lorentz transformation (2.3.15) in full. We then find that

$$t_a = (t'_a + vx'_a/c^2)$$
$$t_b = (t'_b + vx'_b/c^2)$$

(3.3.2)

but $x'_a = x'_b$, (recall figure 3.1) and so

$$t_a - t_b = t = \gamma(t'_a - t'_b) = \gamma t_0$$

Next consider the Lorentz contraction (3.1.7); the statement was made in §3.1 that the observer travelling in the meson reference frame asserts that he has moved a distance vt_0 in the laboratory between the birth and death of the

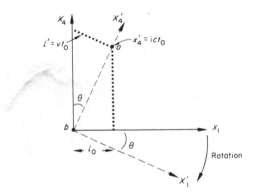

Figure 3.3

meson. Figure 3.3 enables us to reach the same conclusion, since from (2.6.12)

$$x'_1 \tan\theta = ict_0 \cdot -i\frac{v}{c} = vt_0 = l'$$

An inspection of the figure immediately enables us to write (with the aid of (2.6.13))

$$l' \cos\theta = l_0 \quad \text{or} \quad l' = l_0/\gamma = l_0\sqrt{(1-v^2/c^2)}$$

(3.3.3)

which is the equation of the Lorentz contraction (3.1.7). It is a simple matter to obtain the same result by writing out the equations of the Lorentz transformation in full—as we did for time dilatation (3.3.2).

Now consider certain other aspects of the Lorentz transformation—we shall consider time dilatation only, but the results apply equally well to Lorentz contraction. In §2.6 we associated a clockwise rotation of axes with the movement of the Σ' reference frame along the direction of the $+X_1$ axis of the Σ frame. It is a simple matter to show that an anticlockwise rotation then implies motion in the opposite direction; that is, $-v$ instead of $+v$ for the velocity. An inspection of figures 3.2 and 3.4(a) (both have the same scales) then shows that the dilatation equation $t=\gamma t_0$ still holds without further calculation. This result is an example of the isotropy of space.

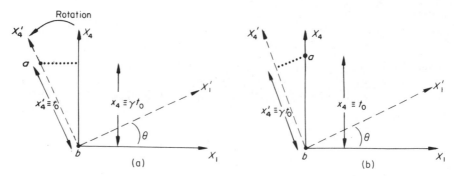

Figure 3.4

Next we show in figure 3.4(b) that which is regarded as the primed and which the unprimed system is irrelevant, provided the appropriate physical conditions are satisfied. A measurement of a time interval t_0 for a system at rest in the Σ frame produces a time $t'=\gamma t_0$ for an observer moving past Σ in a reference frame Σ'.

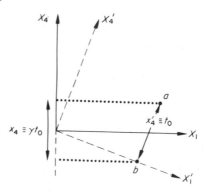

Figure 3.5

Lastly consider figure 3.5. This represents an example of invariance under displacements (we anticipate §5.6 with this diagram). An inspection of this figure shows that x_4 remains unaltered if we displace the spatial coordinates of both a and b from the origin—only the condition $x'_a=x'_b$ is important.

3.4 Simultaneity and causality

Two events are said to be *simultaneous* if an observer records them as both occurring at the same time. Thus an observer located at O in the Σ frame

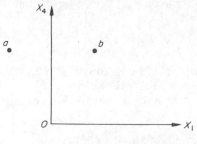

Figure 3.6

(figure 3.6) would describe the events a and b as simultaneous if the line joining them lay parallel to the X_1 axis. Now from a reference frame Σ' an observer would record the time difference between the events as

$$t'_a - t'_b = \gamma [(t_a - t_b) - (v/c^2)(x_a - x_b)] \tag{3.4.1}$$
$$= (\gamma v/c^2)(x_a - x_b) \tag{3.4.2}$$

where we have used the last of equations (2.6.13) and set $t_a = t_b$ because of the simultaneity condition. It is apparent from equation (3.4.2) that the events will only appear to be simultaneous in the Σ' frame if $x_a = x_b$, that is two events must coincide *both in space and time* in order to appear simultaneous in another Lorentz reference frame.

In figures 3.7 we note another aspect of the problem. The direction of rota-

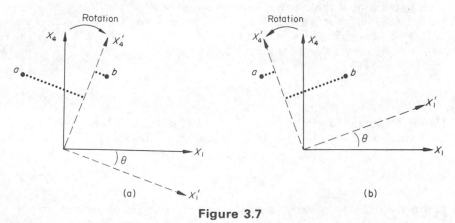

Figure 3.7

tion of axes in these figures is associated with the direction of the relative velocities of the Σ and Σ' frames. An inspection of figure 3.7 shows that event a can be recorded earlier or later than b in the frame Σ' according to the direction of the velocity. This result is also apparent in equation (3.4.2).

Now let us go further and enquire what happens when events are not simultaneous in the Σ frame. We shall first consider causally related events. When discussing the light cone in §2.5 we noted that events at a and b were causally related provided that

$$c(t_a - t_b) > (x_a - x_b) \qquad (3.4.3)$$

that is they must be separated by a distance less than the product of the time interval between the events times the velocity of light. In terms of our $X_1 X_4$ coordinates this condition implies that the line joining a and b must be less

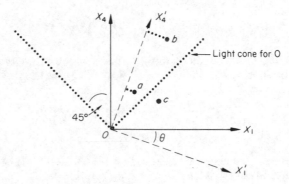

Figure 3.8

than 45° to the X_4 (time) axis — compare our discussion of the light cone in §2.5. In addition if the events are to be causal with respect to an observer located at O in figure 3.8, then a and b must lie within O's light cone—the dotted lines of figure 3.8. Now the angle θ in figure 3.8 is related to the equations of the Lorentz transformation (§2.6)

$$\tan\theta = -\mathrm{i}\,\frac{v}{c}$$

Hence the physical limits in θ are $0° < \theta < 45°$, since $v \leqslant c$. It is immediately apparent from figure 3.8 that for any position of the axis OX'_4 inside the light cone

$$t'_b > t'_a \quad \text{if} \quad t_b > t_a \qquad (3.4.4)$$

that is *the ordering in time of causally related events is not affected by the Lorentz transformation*. No such restrictions need apply to acausally related events—consider for example the points a and c in figure 3.8 when the line OX'_4 is rotated close to the 45° limit, then $t'_a < t'_c$ whilst $t_a > t_c$.

3.5   The clock paradox

In this section and the next we shall examine two oddities which arise in the theory of special relativity. The clock or twin paradox was first pointed out by Einstein in 1905 and was the subject of some controversy in the 1950s.* Stated briefly in the language of space travel, we begin by supposing we have two identical twins, one an astronaut and one a ground controller. The astronaut makes a space flight and returns in T_0 years according to a clock in his reference frame. However his brother records the elapse of time as γT_0 and since $\gamma > 1$ the astronaut will be younger than his brother upon his return.

The 'paradox' lies in the asymmetric nature of the conclusion which appears to contradict the theory of relativity, since one might argue that from the astronaut's rest frame it is his twin who made the journey and therefore the ground controller should be younger when they meet again. We show below that this conclusion cannot be true.

To simplify labelling let us call the astronaut A and his twin the controller C, and to clarify the discussion we shall examine two journeys (figure 3.9—note that this is not a space-time diagram). We shall assume that the times to accelerate to a constant velocity v, reverse, and decelerate are the same for both

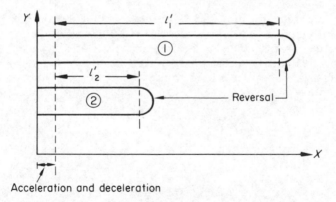

Figure 3.9

journeys and will label them as t_{ARD}. The distances l' are measured in A's frame, then the time for journey 1 is

$$T_1' = \frac{2l_1'}{v} + t_{\mathrm{ARD}}' \text{ in A's frame}$$

$$T_1 = \frac{2\gamma l_1'}{v} + t_{\mathrm{ARD}} \text{ in C's frame}$$

(3.5.1)

---

* Many of the papers appearing during the controversy may be found in 'Selected reprints on special relativity theory', American Institute of Physics, 1963. See also SACHS, M., *Physics Today*, **23**, Sep. 1971.

and for journey 2

$$T_2' = \frac{2l_2'}{v} + t_{ARD}'$$

$$T_2 = \frac{2\gamma l_2'}{v} + t_{ARD}$$

(3.5.2)

hence

$$T_1 - T_2 = \gamma(T_1' - T_2')$$

(3.5.3)

and thus the time which has elapsed for C is longer than A. By considering two journeys we have eliminated the end effects; alternatively we could obtain the same result from a single journey by making the journey very long so that $l_1'/v \gg t_{ARD}'$.

The reason why the situation is not symmetric between A and C lies in the t_{ARD} regions. In these regions the astronaut is only too well aware that he is in a non-inertial frame due to the pressure placed on his body by the non-uniform motion. C experiences no pressures and so the two frames are not equivalent.

If we ignore the problems raised by the nature of the rockets which have to yield (1) acceptable accelerations so as not to kill the astronaut and (2) final velocities close to that of light, then the dilatation effect is most useful in travelling to distant stars, then an astronaut could go on a round trip to a star 999 light years away at a speed of $0.99995c$ and take a time

$$\frac{2 \times 999 \times c}{0.99995c} \text{ years} \sim 2000 \text{ years}$$

in the earth's frame. The astronaut will say the journey involved

$$\frac{2000}{\gamma} = 2000 \sqrt{(1 - 0.99995^2)}$$
$$= 2000 \sqrt{\{1 - (1 - 0.00005)^2\}}$$
$$= 20 \text{ years}$$

and so he will return to earth 20 years older—that is 20 years at the same pace (heartbeats, physical ageing) as if he had stayed on earth.

One experimental check of the clock paradox has been made. This was achieved in the experiment which checked the dilatation factor for muons (§3.1). Consider the situation shown in figure 3.10. Let there be a ground controller C stationed at the origin of an inertial frame Σ and an astronaut A in a frame Σ' executing uniform rotation about C with velocity v as defined in the Σ frame. Because of centrifugal forces A is obviously not in an inertial frame. Applying our standard formulae, we find

(1) times in C and A are related by $t = \gamma t'$

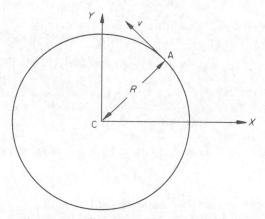

Figure 3.10

(2) the length of the circumference of the circle will be reduced for the observer A in Σ' by the factor $\sqrt{(1-v^2/c^2)}$

(3) the radius R will be the same for both observers.

Thus in addition to the clock problem, we find that the ratio of the circumference of the circle to its diameter is less than π for the observer A. We can therefore see that the problems associated with non-inertial frames do not stop with clocks.

The two experiments, which we mentioned in §3.1 and which acted as checks on the time dilatation formula, involved straight line paths for pions and circular orbits for the muons. Let us consider the latter case in a little more detail.

If we imagined our astronaut to be executing a circular path, and that he carried a source of stationary muons with him the number left after a time t' would be

$$N_A = N \exp(-t'/\tau_0) \qquad (3.5.4)$$

where τ_0 is the mean lifetime of the muons and N the number of muons generated at time $t'=0$, then from conclusion (1) above

$$N_A = N \exp(-t/\gamma\tau_0) \qquad (3.5.5)$$

If his twin brother C also has a source of stationary muons he would find the number of remaining after a time t to be

$$N_C = N \exp(-t/\tau_0)$$

Thus we would expect a discrepancy between N_A and N_C and this discrepancy will depend on γ.

Numerous experiments with muons at rest in the laboratory frame have established the value τ_0 to 2.20×10^{-6} s. Instead of employing astronauts and a spaceship the experiment of the CERN group (§3.1) involved sending the muons around in circular orbits with a momentum of 1.284 GeV/c. They found that the muons decayed according to the relationship

$$N_A = N \exp[-t/(26.37 \pm 0.05) \, 10^{-6}] \tag{3.5.6}$$

Thus the muons 'in orbit' live longer; a muon clock is as good as any other as an indicator of time, and so we conclude that the astronaut would also live longer. To make the experiment quantitative, a momentum of 1.284 GeV/c corresponds to $\gamma = 12.14$ (§4.3), so that $\gamma\tau_0 = 12.14 \times 2.20 \times 10^{-6} = 26.69 \times 10^{-6}$ s. Thus according to equation (3.5.5) we should expect

$$N_A = N_0 \exp[-t/(26.69 \times 10^{-6})] \tag{3.5.7}$$

The agreement between equations (3.5.6) and (3.5.7) is impressive, and it suggests that the rate of the clock (that is, the muon decay rate) does not depend on the acceleration, but only on the velocity.* The discrepancy between the two numbers was interpreted as a loss of muons due to imperfections in the magnetic field.

3.6   The visual appearance of rapidly moving bodies

In his paper of 1905 Einstein states[†] 'A rigid body which, measured in a state of rest, has the form of a sphere, therefore has in a state of motion—viewed from the stationary system—the form of an ellipsoid of revolution with the axes

$$R \sqrt{(1 - v^2/c^2)}, R, R.'$$

He goes on to conclude that for $v \to c$ 'all moving objects—viewed from the 'stationary' system—shrivel up into plain figures.

There has since been some controversy whether Einstein meant 'viewed' to be interpreted as a measurement or a visual impression. As we shall see below there is a difference and the visual appearance can lead to some odd behaviour.[‡]

Let us recall the discussion of the Lorentz contraction (§3.2 and 3.3). Then it was decided that the length l_0 of a rod, as measured in a frame in which the rod was at rest, would be determined to be

$$l' = l_0/\gamma = \sqrt{(1 - v^2/c^2)} \, l_0 \tag{3.6.1}$$

when the measurement was made from a frame moving with relative velocity v.

The determination can be made in a number of ways. Let us assume that we measure a length l'_0 for a rod at rest in the Σ' (moving) frame

$$l'_0 = x'_a - x'_b$$

---

* The acceleration experienced by the muons was high; the modulus of the four-acceleration (§4.9) was $\sim 5 \times 10^{19}$ m/s^2.
† Translation by W. Perrett and G. B. Jeffrey, *The Principle of Relativity*, Methuen, 1923.
‡ A recent paper giving many references on this topic is that of SCOTT, G. D., and VAN DRELL, H. J., *Am. J. Phys.* **38**, 971, 1970.

where the coordinates x'_a and x'_b record the positions of the ends of the rod. The Lorentz transformation (2.3.15), with $t_a = t_b$

$$x'_a = \gamma (x_a - vt_a) \qquad x'_b = \gamma (x_b - vt_b) \tag{3.6.2}$$

then gives the length of the rod in the Σ frame

$$l'_0 = x'_a - x'_b = \gamma (x_a - x_b) = \gamma l \tag{3.6.3}$$

or

$$l = \sqrt{(1 - v^2/c^2)}\, l'_0 \tag{3.6.4}$$

Notice that the dashes have changed side between equations (3.6.1) and (3.6.4). This is because the length l_0 was determined for a rod at rest in the Σ frame for (3.6.1). The presence of the contraction factor $\sqrt{(1 - v^2/c^2)}$ illustrates the principle of relativity, namely that no one reference frame is preferred over another.

Notice also that equations (3.6.2) and (3.6.3) imply that the measurements x_a and x_b must be carried out *simultaneously* in the Σ frame. The connection is acausal (§3.4) and so two separate observers are necessary to establish the Lorentz contraction. A single observer can also determine the Lorentz contraction by measuring the time the moving rod takes to pass him in the frame Σ; however, this consideration is irrelevant, since what we are after is what an observer sees, not what he measures.

An eye (or a camera) of a single observer records an image at a given spatial point and at a given instant of time. When account is taken of the finite velocity of light, it is apparent that the light received by the eye has not been emitted simultaneously by all points of the object. The points furthest away emitted their light earlier than the closest points. Thus, if a body is moving, the picture received by the eye will be different from that received if the body is at rest, since the coordinates of the body will have changed during the time the various points on it emitted the light received by the observer.

Consider light being emitted from a world point a. As we have already seen in §2.3, the light front forms a spherical shell in both Σ and Σ' frames and propagates according to the relations

$$
\begin{aligned}
(x' - x'_a)^2 + (y' - y'_a)^2 + (z' - z'_a)^2 &= c^2 (t' - t'_a)^2 \text{ in } \Sigma' \\
(x - x_a)^2 + (y - y_a)^2 + (z - z_a)^2 &= c^2 (t - t_a)^2 \text{ in } \Sigma
\end{aligned} \tag{3.6.5}
$$

Now for an observer located at the spatial origin of the Σ frame, that is $x = y = z = 0$ the light front satisfies the following equation at a time $t = t_0$.*

$$x_a^2 + y_a^2 + z_a^2 = c^2 (t_0 - t_a)^2. \tag{3.6.6}$$

---

* We follow the approach of S. YNGSTRÖM, *Arkiv för Fysik*, **23**, 367, 1965. Once again we assume that the Σ' frame moves in the X direction with a velocity v and that the X and X' axes are collinear.

Thus this equation defines the locus of the light sources from which light has started at times t_a, so defined that their wave fronts arrive simultaneously to the observer at a time $t = t_0$. This locus represents the apparent position of a at time t_0. Since $t_0 > t_a$ (that is, later in time than t_a according to causality) equation (3.6.6) becomes

$$\sqrt{(x_a^2 + y_a^2 + z_a^2)} = c\,(t_0 - t_a) \tag{3.6.7}$$

and upon inserting our standard equation of the Lorentz transformation (2.3.15)

$$x' = \gamma\,(x - vt) \rightarrow t_a = (1/v)\,(x_a - x_a'/\gamma)$$

we find

$$(v/c)\sqrt{(x_a^2 + y_a^2 + z_a^2)} = (vt_0 + x_a'/\gamma - x_a) \tag{3.6.8}$$

Without loss of generality we can take $t_0 = 0$ and if we write

$$R_a = \sqrt{(x_a^2 + y_a^2 + z_a^2)}$$

and recall that the y_a' and z_a' coordinates remain unchanged for a Lorentz transformation in the X direction, then we have (with $\beta = v/c$)

$$\begin{aligned} x_a' &= \gamma\,(x_a + \beta R_a) \\ y_a' &= y_a \\ z_a' &= z_a \end{aligned} \tag{3.6.9}$$

These equations relate coordinates in the Σ and Σ' frames at a fixed time of observation. As an example of their application consider the case of a thin rod of length l_0 as measured in the Σ' frame. For simplicity we shall assume that the rod lies in the $X'Y'$ plane so that $z = z' = 0$ and that it is parallel to the X' axis (figure 3.11).

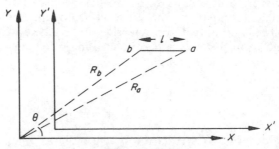

Figure 3.11

The length of the rod is given by

$$l_0 = x_a' - x_b'$$

in the Σ' frame. Thus from equations (3.6.9) the observer in the Σ frame 'sees' the length as

$$l=(x_a-x_b)=(1/\gamma)\,(x_a'-x_b')-\beta\,(R_a-R_b)$$
$$=(l_0'/\gamma)-\beta\,(R_a-R_b) \tag{3.6.10}$$

The cosine theorem gives us

$$R_a^2=R_b^2+l^2-2R_bl\,\cos(\pi-\theta)$$

or

$$R_a-R_b=R_b\,\{\sqrt{[1+(l/R_b)^2+(2l/R_b)\,\cos\theta]}-1\}$$

If l/R_b is a small quantity the use of the binomial expansion reduces the right-hand side of the above equation to $l\cos\theta$ and so equation (3.6.10) yields the result

$$l=\frac{l_0'\sqrt{(1-v^2/c^2)}}{1+(v/c)\,\cos\theta} \tag{3.6.11}$$

Thus only at $\theta=90°$ does the observer 'see' the normal Lorentz contraction. If we consider the situations when $\theta\to\pi$ and 0, which corresponds, respectively to the situations when the rod is approaching from the left and receding to the right (and is at large distances from the observer) the result can be represented by figure 3.12. The conventional Lorentz contraction is represented by the broken line in this figure.*

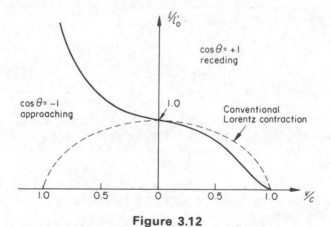

Figure 3.12

The calculation of the appearance of more sophisticated shapes tends to be rather lengthy despite the apparent simplicity of equations (3.6.9). The numerical evaluation is therefore normally performed by a computer, and in figure

---

* The result displayed in this figure was first obtained by R. WEINSTEIN, *Am. J. Phys.* **28**, 607, 1960.

3.13 we display the results of one such evaluation.* In this figure we see the appearance of a moving grid as seen by an observer placed at unit distance

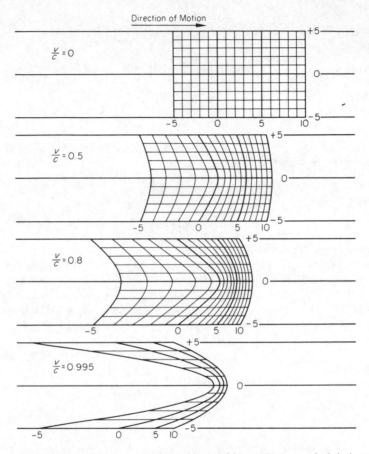

Figure 3.13 The appearance of a plane grid moving at relativistic speeds. The observer is unit distance in front of the origin. For each view the direction in which the observer sees the origin is perpendicular to the motion.

from the origin along the Z axis when the grid lies in the XY plane and moves in the X direction. The distortion associated with equation (3.6.11) and figure 3.12 is readily apparent.

We return now to Einstein's remark concerning spheres which opened this section. Whilst no one doubts that his remark is perfectly true within the context of physical measurement, nevertheless the appearance of the sphere to a single observer will depend on the physical conditions. Detailed analysis[†],

* SCOTT, G. D., and VINER, M. R., *Am. J. Phys.* **33**, 534, 1964.
† BOAS, M. L., *Am. J. Phys.* **29**, 283, 1961.

using the technique described above, shows that a sphere always presents a circular disc in outline to the observer. However any shapes appearing on the surface of the sphere become distorted as $v \to c$.

Finally we mention that it is not only the shape of rapidly moving objects which changes. In §4.6 we shall show that both the colour and brightness of a rapidly moving object change as it approaches an observer and then recedes.

3.7   The relativistic transformation of velocities

So far, we have considered the relation between spatial and time coordinates according to the Lorentz transformation. We shall now examine the first-order differentials of these terms—the velocities.

In §1.1 it was shown that if a body moved with velocity u' along the X' axis in the Σ' frame, and if the Σ' frame possessed a velocity v relative to the X axis of the Σ frame, then the velocity determined by an observer in the Σ frame is

$$u' = u - v$$

according to the Galilean transformation.

For our present purposes we shall have to examine velocities in the X, Y and Z directions; however, as the results for Y and Z are symmetric (apart from a relabelling of subscripts) we shall merely quote the results for the Z axis and leave the details as a problem to the reader.

Let us introduce velocities

$$u'_x = \frac{dx'}{dt'} \qquad u'_y = \frac{dy'}{dt'}$$

in a reference frame Σ' which moves with a velocity v along the X axis of the Σ reference frame. Then application of the equations of the Lorentz transformation (2.3.15) give us

$$
\begin{aligned}
dx &= \gamma(dx' + v\,dt') &= \gamma(u'_x + v)\,dt' \\
dy &= dy' &= u'_y\,dt' \\
dt &= \gamma(dt' + v\,dx'/c^2) = \gamma(1 + vu'_x/c^2)\,dt'
\end{aligned}
\qquad (3.7.1)
$$

hence

$$
\begin{aligned}
u_x &= \frac{dx}{dt} = \frac{u'_x + v}{1 + vu'_x/c^2} & u'_x &= \frac{u_x - v}{1 - vu_x/c^2} \\
u_y &= \frac{dy}{dt} = \frac{u'_y}{\gamma(1 + vu'_x/c^2)} & u'_y &= \frac{u_y}{\gamma(1 - vu_x/c^2)} \\
u_z &= \frac{dz}{dt} = \frac{u'_z}{\gamma(1 + vu'_x/c^2)} & u'_z &= \frac{u_z}{\gamma(1 - vu'_x/c^2)}
\end{aligned}
\qquad (3.7.2)
$$

An inspection of the above equation shows that in the non-relativistic limits $u_x \to 0$ and $v \to 0$, then

$$u'_x = u_x - v$$

which, apart from the subscripts, is the equation obtained from the Galilean transformation (1.1.3). At the other extreme in the limits $u_x \to c$ and $v \to c$ we find $u'_x \to c$. Thus the maximum attainable velocity in any reference frame is that of light. As we indicated in §2.2 this equation has been verified to a high order of accuracy by using γ-rays from the decay of π^0 mesons.

An interesting aspect of equations (3.7.2) is the behaviour of the y and z components of the velocity. Unlike the y and z components of coordinates (2.3.15), they do not remain invariant under Lorentz transformations in the X direction. The difference arises because of the behaviour of the time intervals in the Lorentz transformation.

The relation between the x components of the velocity can also be easily obtained by considering a rotation of axes in space-time (§2.6). Consider a particle which leaves the origin, O', of the Σ' frame at a time $t' = 0$ (figure 3.14) and assume that the origins of the Σ and Σ' frames coincide at this time. If the

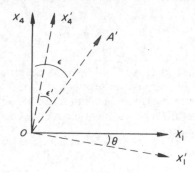

Figure 3.14

particle moves with a uniform velocity u'_x in the Σ' frame then its path as a function of time is given by OA' where (compare (2.6.12))

$$\tan \varepsilon' = \frac{x'_1}{x'_4} = \frac{x'_1}{ict'} = \frac{-iu'_x}{c} \tag{3.7.3}$$

Similarly in the Σ frame

$$\tan \varepsilon = -i\frac{u_x}{c} \tag{3.7.4}$$

and the two frames are related by (2.6.12)

$$\tan \theta = -i\frac{v}{c} \tag{3.7.5}$$

where v represents the velocity of the Σ' frame with respect to the Σ frame.

The law for the addition of tangents

$$\varepsilon = \varepsilon' + \theta$$
$$\tan \varepsilon = \frac{\tan \varepsilon' + \tan \theta}{1 - \tan \varepsilon' \tan \theta} \tag{3.7.6}$$

then gives us

$$u_x = \frac{u'_x + v}{1 + u'_x v/c^2}$$

which is the first of equations (3.7.2). The transformation of velocities in the Y and Z directions can also be obtained geometrically but spheres are then necessary and the construction is not so obvious.

3.8   Successive Lorentz transformations

The geometrical treatment of the transformation of velocities can also be used to consider two successive Lorentz transformations. Consider three reference frames Σ, Σ' and Σ'', and let the Σ' frame move with velocity v along the X axis of the Σ frame; then from the equations of the Lorentz transformation (2.3.15)

$$\begin{aligned} x' &= \gamma (x - vt) \\ t' &= \gamma (t - vx/c^2) \end{aligned} \tag{3.8.1}$$

Similarly if the Σ'' frame moves with a velocity v' with respect to Σ' along the X' axis, that is, the X axis, then

$$\begin{aligned} x'' &= \gamma' (x' - v't') \\ t'' &= \gamma' (t' - v'x'/c^2) \end{aligned} \tag{3.8.2}$$

where

$$\gamma' = \frac{1}{\sqrt{(1 - v'^2/c^2)}}$$

Obviously it is possible to eliminate x' and t' from the above equations and obtain a relation between x'', t'' and x, t. However, the result can be obtained immediately by considering a double rotation of axes (figure 3.15). A comparison of this figure with that of figure 2.6, and an inspection of equations (2.6.11) permit us to write

$$\begin{aligned} x''_1 &= x_1 \cos \theta'' - x_4 \sin \theta'' \\ x''_4 &= x_1 \sin \theta'' + x_4 \cos \theta'' \end{aligned} \tag{3.8.3}$$

where

$$\tan \theta'' = -iv''/c$$

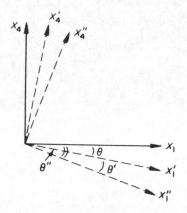

Figure 3.15

and v'' represents the velocity of the Σ'' reference frame with respect to the Σ frame (compare (3.7.2) and (3.7.6).)

$$v'' = \frac{v' + v}{1 + v'v/c^2}$$

Thus the product of two Lorentz transformations is itself a Lorentz transformation, provided that both transformations are made in the same direction. If this condition, is not fulfilled the transformation is more complicated; we shall raise this point again in §4.5.

PROBLEMS

3.1 The mean lifetime of a muon in its own rest frame is 2.0×10^{-6} s. What average distance would the particle travel in vacuum before decaying when measured in reference frames in which its velocity is $0c, 0.6c, 0.99c$? Determine also the distances through which the muon claims it travelled.

3.2 Consider the following idealised clock. It consists of two mirrors between which a light pulse bounces forwards and backwards, and a dial which records one count for each return of the pulse. Let the clock move transversely to its length at a velocity v in a reference frame Σ. Show that if the intervals between successive counts is T_0 in the clock's frame, then it is γT_0 in the Σ frame.

3.3 Two standard clocks A and B move freely through space with constant velocities. They meet momentarily, when both are set to zero. Later upon registering a time T_A, clock A emits a light pulse which is received by clock B and B registers T_B. Show that the velocity of one clock in the reference frame of the other is given by $c(T_B^2 - T_A^2)/(T_B^2 + T_A^2)$.

3.4 Show that the Lorentz contraction equation (3.3.3) may be obtained from the basic equations of the Lorentz transformation (2.3.15) by making an acausal measurement on the ends of a rod (that is, at the same time at different places). Show that the same result can also be obtained by a single observer by a causal measurement, if he records the time interval for the rod to pass a fixed point.

3.5 Two rocket ships A and B pass each other. The length of A in its own reference frame is l_0 and B determines the same length to be l_0/γ. Show that A's determination of B's measurement is l_0.

3.6 Two events A and B are simultaneous in a reference frame Σ in which they are spatially separated by a distance X. Show that in all other frames Σ', moving relative to Σ along the line joining the events, the measured spatial separation is always greater than X. How do you reconcile this result with the usual concept of length contraction?

3.7 On Jan 1st 1980 a rocket ship passes the earth at a speed of $0.8c$ travelling to the nearest star, α-Centauri, which is 4 light years distant in the earth's reference frame. As this rocket ship reaches the star a second ship B, heading for the earth at a speed $0.8c$ relative to the earth, passes A and the observer in B sets his clocks and calendar identical to that of A. What is the date measured on earth and in the ship B when B passes earth?

3.8 In the previous question what are the dates on earth as determined by observers in the rocket ships A and B at the moment when they pass?

3.9 Consider a reference frame Σ with an observer located at a considerable distance from the origin along the Y axis. A small cube with sides parallel to the X axis moves along the X axis with velocity v. Show that when the cube is near $x=0$ it is apparently rotated by an angle θ, given by $\sin\theta = v/c$, to the observer located on the Y axis?

3.10 In question 1.2 the problem was considered of the relative velocity of two electrons which each move with velocity $0.8c$ in opposite direction in the laboratory. In that question you were asked to determine the relative velocity of one electron in the frame of the other by means of a Galilean transformation; what answer do you get for a Lorentz transformation?

4

Four-vectors

4.1   Introduction

Up to now we have avoided using the word vector in this book. In introductory textbooks in physics, a vector is normally defined as something which possesses both magnitude and direction in space. It is also normally printed in heavy type, for example the velocity u of a body is written as *

$$\mathbf{u} = \mathbf{i}u_x + \mathbf{j}u_y + \mathbf{k}u_z \tag{4.1.1}$$

where $\mathbf{i}, \mathbf{j}, \mathbf{k}$ are unit vectors pointing along the X, Y, Z axes respectively and u_x, u_y, u_z represent the components of u along these axes. Some of the definitions used in vector notation are given in appendix 2.

If the reference axes are rotated, the components of $\mathbf{u}$ along the axes will change but the magnitude of $\mathbf{u}$ remains unaltered (compare OP in figure 2.5).

---

* We shall occasionally use the symbol $|\mathbf{u}|$ to denote the magnitude of $\mathbf{u}$, thus
$$|\mathbf{u}| = \sqrt{(u_x^2 + u_y^2 + u_z^2)} = u.$$

In a similar fashion we may introduce four-vectors, whose magnitudes remain unaltered under rotations in space-time. As we have seen in chapter 2, these rotations may be identified with the Lorentz transformations.

We shall therefore define a four-vector A as any system possessing components A_1, A_2, A_3, A_4 which behave like the space-time coordinates x_1, x_2, x_3, x_4 under Lorentz transformations (2.6.14)

$$
\begin{aligned}
A'_1 &= \gamma A_1 + i\beta\gamma A_4 \\
A'_2 &= A_2 \\
A'_3 &= A_3 \\
A'_4 &= -i\beta\gamma A_1 + \gamma A_4
\end{aligned}
\tag{4.1.2}
$$

One example of a four-vector is, of course, the space-time coordinate frame x_1, x_2, x_3, x_4. We shall encounter others in this chapter.

The scalar product of three-dimensional vectors is normally written as

$$\mathbf{u} \cdot \mathbf{v} = u_x u_x + u_y v_y + u_z v_z \tag{4.1.3}$$

where the dot indicates the cosine of the angle between the two vectors. When dealing with four-vectors we shall denote a scalar product by an expression of the type

$$\sum_{\lambda=1}^{4} A_\lambda B_\lambda = A_1 B_1 + A_2 B_2 + A_3 B_3 + A_4 B_4 \tag{4.1.4}$$

It is a simple matter to show that the scalar product remains invariant under a common Lorentz transformation

$$\sum_{\lambda=1}^{4} A'_\lambda B'_\lambda = \sum_{\lambda=1}^{4} A_\lambda B_\lambda \tag{4.1.5}$$

It is a common practice to drop the summation sign when repeated indices occur. We shall retain them for greater clarity.

One example of a scalar product is the interval (§2.4). In this case the product is that of a four-vector with itself.

Two other points are worth introducing at this stage. Firstly we shall occasionally use a form of shorthand and write a four-vector as

$$A = (\mathbf{A}, A_4) \tag{4.1.6}$$

where $\mathbf{A}$ represents the three spatial components. The scalar product of two four-vectors then appears as

$$\sum_{\lambda=1}^{4} A_\lambda B_\lambda = \mathbf{A} \cdot \mathbf{B} + A_4 B_4 \tag{4.1.7}$$

When the scalar product is that of a four-vector with itself we shall represent it as

$$\sum_{\lambda=1}^{4} A_\lambda A_\lambda = \sum_\lambda A_\lambda^2 = \mathbf{A} \cdot \mathbf{A} + A_4^2 = \mathbf{A}^2 + A_4^2 \qquad (4.1.8)$$

The second point is also one of notation. It will occasionally be convenient to use matrix algebra. We therefore introduce a 4×4 matrix L with components $L_{\lambda\alpha}$, where the indices λ, α run from 1 to 4

$$L = \begin{array}{c} \\ \end{array} \begin{array}{cccc} & \lambda \rightarrow & & \\ 1 & 2 & 3 & 4 \\ \end{array}$$

$$L = \begin{bmatrix} \gamma & 0 & 0 & i\beta\gamma \\ 0 & 1 & 0 & 0 \\ 0 & 0 & 1 & 0 \\ -i\beta\gamma & 0 & 0 & \gamma \end{bmatrix} \begin{array}{c} 1\ \alpha \\ 2\ \downarrow \\ 3 \\ 4 \end{array} \qquad (4.1.9)$$

The symbol L represents the Lorentz transformation operator, and we can display all the equations (4.1.2) by the single equation

$$A'_\lambda = \sum_{\alpha=1}^{4} L_{\lambda\alpha} A_\alpha \qquad (4.1.10)$$

For example if $\lambda = 1$ we have

$$\begin{aligned} A'_1 &= L_{11}A_1 + L_{12}A_2 + L_{13}A_3 + L_{14}A_4 \\ &= \gamma A_1 + i\beta\gamma A_4 \end{aligned} \qquad (4.1.11)$$

which is the first of equations (4.1.2). An inspection of the above equation and (4.1.9) in conjunction with (2.6.15) shows that inverse of the transformation can be represented as

$$A_\lambda = \sum_{\alpha=1}^{4} L_{\alpha\lambda} A'_\alpha \qquad (4.1.12)$$

for example if $\lambda = 1$ then from (4.1.9)

$$\begin{aligned} A_1 &= L_{11}A'_1 + L_{21}A'_2 + L_{31}A'_3 + L_{41}A'_4 \\ &= \gamma A'_1 - i\beta\gamma A'_4 \end{aligned} \qquad (4.1.13)$$

This is the top right equation in (2.6.15) if A is chosen to represent the four-vector x.

4.2   The energy–momentum four-vector

One of the most important four-vectors used in the physics of elementary particles is that for energy-momentum. Let us recall the expression for the interval between two events a and b (2.6.2)

$$-S_{ab}^2 = x_1^2 + x_2^2 + x_3^2 + x_4^2 = \sum_{\lambda=1}^{4} x_\lambda^2$$

If the two events are connected by the passage of a particle with velocity u, then for an element of the interval

$$dS^2 = -\sum_\lambda dx_\lambda^2 = c^2\, dt^2 - (dx_1^2 + dx_2^2 + dx_3^2)$$

$$= c^2\, dt^2 \left(1 - \frac{dx_1^2 + dx_2^2 + dx_3^2}{c^2\, dt^2}\right) \tag{4.2.1}$$

but

$$dx_1^2 + dx_2^2 + dx_3^2 = u^2\, dt^2$$

and so

$$dS^2 = c^2\, dt^2\,(1 - u^2/c^2) = (c^2/\gamma_u^2\, dt^2) = c^2\, d\tau^2 \tag{4.2.2}$$

where

$$\gamma_u = \frac{1}{\sqrt{(1 - u^2/c^2)}}, \qquad d\tau = \frac{dt}{\gamma_u}$$

and the time interval $d\tau$ is called the Lorentz invariant element of time since S and c are Lorentz invariants.

Now dS is a Lorentz invariant quantity, and so if we introduce a function

$$U_\lambda = c\,\frac{dx_\lambda}{dS} = \gamma_u\,\frac{dx_\lambda}{dt} = \frac{dx_\lambda}{d\tau} \tag{4.2.3}$$

it will transform in the same manner as dx_λ, so that, in the spirit of equation (4.1.10), we can write

$$U'_\lambda = \sum_\alpha L_{\lambda\alpha} U_\alpha \tag{4.2.4}$$

Furthermore from (4.2.1) the scalar product of U_λ with itself is

$$\sum_\lambda U_\lambda^2 = c^2 \sum_\lambda \frac{dx_\lambda^2}{dS^2} = -c^2 \tag{4.2.5}$$

and so $\sum_\lambda U_\lambda^2$ remains invariant under Lorentz transformations.*

Thus U_λ is a four-vector, and it is called a *four-velocity*, since from (4.2.3) its components are

$$U_\lambda = \gamma_u\,\frac{dx_\lambda}{dt}$$

$$= \gamma_u\,(u_1, u_2, u_3, ic) \tag{4.2.6}$$

---

* Expressions which remain invariant under transformations are often called *scalars*, for example $\sum_\lambda U_\lambda^2$ is a Lorentz scalar.

and in the nonrelativistic limit, $\gamma_u \to 1$, the spatial components of U_λ are recognisable as the components of the conventional velocity of a particle

$$U_j \to \frac{dx_j}{dt} = u_j$$
$$\qquad\qquad (j = 1, 2, 3)*$$
$$U_4 \to \frac{dx_4}{dt} = ic$$

In nonrelativistic classical mechanics the momentum of a particle with inertial mass m_0 and velocity u is defined as $m_0 u$. Using the definition of four-velocity let us construct a *four-momentum P* with components

$$P_\lambda = m_0 U_\lambda; \quad P = (\mathbf{p}, p_4) = \mathbf{p}, \, i\frac{E}{c} \qquad (4.2.7)$$

then from (4.2.3) and (4.2.6)

$$p_j = \gamma_u m_0 \frac{dx_j}{dt} = \frac{m_0 u_j}{\sqrt{(1 - u^2/c^2)}} \quad (j = 1, 2, 3)$$

$$p_4 = \gamma_u m_0 \frac{dx_4}{dt} = \frac{i}{c} \frac{m_0 c^2}{\sqrt{(1 - u^2/c^2)}} = i\frac{E}{c}$$

$$(4.2.8)$$

An inspection of the above equations shows that the components of the velocity u are given by

$$u_j = \beta_j c = \frac{p_j c^2}{E} \qquad (4.2.9)$$

and, following (4.1.1), the vector **u** *is*

$$\mathbf{u} = \boldsymbol{\beta} c = \frac{\mathbf{p} c^2}{E} \qquad (4.2.10)$$

Consider the first of the equations (4.2.8). The classical expression for the cartesian components of momentum $m_0 u_j$ has become $m_0 u_j/\sqrt{(1 - u^2/c^2)}$. Thus our momentum terms now contain an effective mass

$$m = \frac{m_0}{\sqrt{(1 - u^2/c^2)}} \qquad (4.2.11)$$

which reduces to m_0 in the limit $u/c \to 0$. The term m_0 is called the *rest mass* of

* We shall use Latin indices for spatial components. In order to avoid confusion between the four-velocity and the normal velocity, we shall always use capitals for the four-velocity, for example

$$\mathbf{U} = \gamma_u \frac{d\mathbf{x}}{dt} = \gamma_u \mathbf{u}$$

the body. The equation (4.2.9) has been checked by bending electrons in magnetic fields; in such situations the radius of curvature R of the path of an electron in a field of strength $|\mathbf{B}| = B$ is given by

$$R = \frac{p}{eB} = \frac{mu}{eB} = \frac{m_0}{\sqrt{(1 - u^2/c^2)}} \frac{u}{eB} \qquad (4.2.12)$$

where B is at right angles to p and e is the electronic charge (compare §6.5.2). The velocity u was measured by various techniques and hence m determined. The results of a number of early experiments are shown in figure 4.1. (It is of interest to note that the first published paper commenting on Einstein's

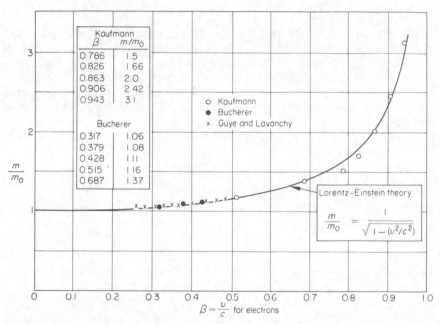

Figure 4.1 Variation of inertial mass with velocity for electrons. Based on the data of Bucherer (1909), Kaufmann (1910), and Guye and Lavanchy (1915). The curve is $1/\sqrt{(1 - u^2/c^2)} = \gamma_u$. (From data assembled by Shankland, R. S., *Atomic and Nuclear Physics, Macmillan, 1961*)

paper of 1905 reported an experiment to check equation (4.2.11). The experiment failed to detect the relativistic increase in mass and so the author stated that Einstein's theory was wrong!)

An inspection of equation (4.2.12) shows that a much higher magnetic field is required to maintain a radius of curvature R than that expected from the classical momentum, $m_0 u$, when $u \rightarrow c$. This effect must always be allowed for in the design considerations for high-energy particle accelerators and their beam lines.

Let us next examine the implications of the term

$$E = \frac{m_0 c^2}{\sqrt{(1 - u^2/c^2)}}$$

in the second of equations (4.2.8). Firstly it has the dimensions of energy, and E is called the *total energy*. Secondly if we consider the nonrelativistic limit and make a binomial expansion, retaining only the first two terms, we find that

$$E = m_0 c^2 (1 + \tfrac{1}{2} u^2/c^2) = m_0 c^2 + \tfrac{1}{2} m_0 u^2 \qquad (4.2.13)$$

Thus, in the nonrelativistic limit, E consists of the classical kinetic energy, $\tfrac{1}{2} m_0 u^2$, plus a term $m_0 c^2$ — this is called the *rest mass energy*.

The appearance of the rest energy is by far the most spectacular and practical consequence of the theory of special relativity. The connection of energy and inertial mass by the proportional constant c^2 implies the release of vast amounts of energy for the destruction of relatively small quantities of matter and vice versa. The destruction of 1 kilogram of matter produces $\sim 9 \times 10^{16}$ joules of energy*; this fact is exploited in the nuclear power stations, and, more horrifically, the destruction of a few grams of matter in a fraction of a second produces a nuclear bomb equivalent to ~ 50 kilotons of TNT in destructive power.

4.3 Further properties of the momentum four-vector

The scalar product of the components of the four-velocity (4.2.5),

$$\sum_\lambda U_\lambda^2 = -c^2$$

implies the existence of following relationship for the momentum four-vector

$$\sum_\lambda p_\lambda^2 = m_0^2 \sum_\lambda U_\lambda^2 = -m_0^2 c^2 \qquad (4.3.1)$$

From (4.2.8) we may expand this equation into its components

$$p_1^2 + p_2^2 + p_3^2 - \frac{E^2}{c^2} = \mathbf{p}^2 - \frac{E^2}{c^2} = -m_0^2 c^2$$

or

$$\boxed{E^2 = \mathbf{p}^2 c^2 + m_0^2 c^4} \qquad (4.3.2)$$

This is the famous Einstein equation relating energy, momentum and mass.

---

* Despite the enormity of this figure the sun is at present destroying itself at the rate of $\sim 5 \times 10^6$ ton/s. Fortunately this rate of loss represents about 1 part in 10^{13} of the sun's mass per year.

Before going further with this equation we shall introduce units which we shall use in future from time to time. Rather than use the MKS system a more convenient unit in high-energy physics is the MeV. This is the energy acquired by a particle of elementary charge e in passing through a potential difference of one million volts; in some circumstances this unit is too small and the GeV is used—the GeV is 10^3 MeV.* An inspection of equation (4.3.2) then shows the relation between E, $\mathbf{p}$ and m_0 to be of the form given in table 4.1.

Table 4.1

Energy	Momentum	Mass
E	$\mathbf{p}$	m_0
MeV	MeV/c	MeV/c^2
GeV	GeV/c	GeV/c^2

Let us return now to equation (4.3.1)

$$\sum_{\lambda=1}^{4} p_\lambda^2 = \mathbf{p}^2 - \frac{E^2}{c^2} = -m_0^2 c^2$$

or

$$\frac{E^2}{c^2} - \mathbf{p}^2 = m_0^2 c^2$$

Now the rest mass of a body is a real quantity; thus the smallest value that the right hand of the last equation can take is zero and so

$$\frac{E^2}{c^2} > \mathbf{p}^2 \qquad (4.3.3)$$

Apart from changes in notation the above condition is the same as that introduced in §2.4 for time-like intervals (compare (2.4.2) and (2.4.3)). The momentum four-vector is thus a time-like four-vector.

Let us next examine the relation between E, p and m_0 in equation (4.3.2)

$$E^2 = \mathbf{p}^2 c^2 + m_0^2 c^4$$

When $\mathbf{p} = 0$ we find that $E = m_0 c^2$ (we choose the positive sign from our definition of E in (4.2.8)). If we make a plot of E against $|\mathbf{p}|\,c$ we can represent this point by A on figure 4.2. From this point a parabola can be constructed according to equation (4.3.3), and it is represented by the broken line in this figure. Let us examine the asymptotic behaviour of this line in the limit $u \to c$; we then find, with aid of (4.3.2) and (4.2.8),

$$E = |\mathbf{p}|\,c\,\sqrt{(1 + m_0^2 c^4/|\mathbf{p}|^2 c^2)} = |\mathbf{p}|\,c\,\sqrt{[1 + (c^2/u^2)\,(1 - u^2/c^2)]}$$
$$\to |\mathbf{p}|\,c \quad \text{for} \quad u \to c. \qquad (4.3.4)$$

---

* In the USA the unit of 10^3 MeV is also called a BeV.

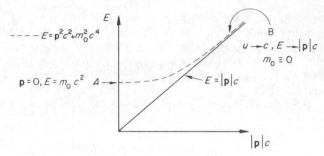

Figure 4.2

Thus in the asymptotic limit the curve of E versus $|\mathbf{p}|c$ is a straight line of slope 45°; this region is indicated by B on figure 4.2. In this region the mass m_0 has no influence on the equation.

$$E^2 = \mathbf{p}^2 c^2 + m_0^2 c^4$$

and so is effectively zero.

Now consider the situation when $u = c$. For particles of finite rest-mass energy and momentum are then infinite. However, the photons of light, which travel with velocity $u = c$, patently do not possess infinite energy; if they also possess zero mass then no problem arises since to construct energies as

$$E = \frac{m_0 c^2}{\sqrt{(1 - u^2/c^2)}}$$

is then a meaningless operation. The construction of the Einstein energy equation

$$E^2 = \mathbf{p}^2 c^2 + m_0^2 c^4$$

involved no specific assumptions about the value of the mass and so it is valid for all masses between zero and infinity. The photons (and also the neutrinos associated with radioactive beta decay) merely represent the lower limit of the range of possible masses, and their relationship between E and $|\mathbf{p}|c$ is displayed by the solid line in figure 4.2.

Instead of using E and $|\mathbf{p}|c$ the data on figure 4.2 could have been displayed by using E and any two of the components $p_x c$, $p_y c$, and $p_z c$ with the third zero. A parabolic surface would then be formed, which in the limit $m_0 \to 0$ would be a cone.

4.4 The Lorentz transformation of the momentum four-vector

Consider a body with rest mass m_0 at rest in a reference frame (figure 4.3 (a)). If we wish to make a Lorentz transformation so that the body appears to move with a velocity u along the X axis to an observer stationed in the Σ frame, we can do it in one of two ways.

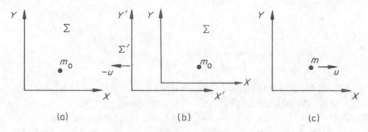

Figure 4.3

(1) Make a Lorentz transformation of the reference frame Σ and the associated observer in the $-X$ direction, that is, create a frame Σ' moving with velocity $-u$ with respect to Σ (figure 4.2 (b)).

(2) Lorentz transform the four-momentum vector for the body so that it moves with velocity u along the $+X$ direction (figure 4.2 (c)).

Both Lorentz transformations are equivalent processes, as we shall indicate below. The process of transforming the coordinate frame (figure 4.3 (b)) is called a *passive transformation*, whilst that of figure 4.3 (c) is called an *active transformation* — it is also sometimes called a *Lorentz boost*.

Let us consider the transformations with the aid of four-vectors and rotations. The momentum four-vector (compare (4.2.8)) for the body at rest is given by

$$p_\alpha = 0, 0, 0, \frac{i}{c} m_0 c^2 \qquad (4.4.1)$$

where the first three zeros refer to the x, y and z components. Since we are considering a Lorentz transformation in the X direction, we can ignore the y and z components for the time being. The equations of the Lorentz transformation for four-vectors (4.1.10), then yield the result

$$p_1' = p_x' = L_{11}p_1 + L_{12}p_2 + L_{13}p_3 + L_{14}p_4$$

$$= i\beta\gamma \cdot \frac{i}{c} m_0 c^2$$

$$= -\gamma m_0 u$$

$$p_4' = \frac{iE'}{c} = L_{41}p_1 + L_{42}p_2 + L_{43}p_3 + L_{44}p_4 \qquad (4.4.2)$$

$$= \gamma \cdot \frac{i}{c} m_0 c^2$$

$$= \frac{i}{c} \gamma m_0 c^2$$

where

$$\beta = \frac{u}{c}; \quad \gamma = \frac{1}{\sqrt{(1 - u^2/c^2)}}$$

Thus

$$p'_\lambda = -\gamma m_0 u, 0, 0, \frac{i}{c}\gamma m_0 c^2$$

which represents a particle moving with velocity $-u$ in the X direction. Thus we have apparently achieved the wrong result, since we wanted our body to move with velocity $+u$. The reason for the discrepancy in sign is simple, but illustrates the dangers of a blind application of formulae. As we stated earlier in this section, the necessary condition for the Lorentz transformation is that it causes the reference frame Σ' to move with velocity $-u$ with respect to X in the Σ frame. Now the matrix (4.19) represents a transformation which causes Σ' to move with velocity $+u$ in the X direction. We must therefore reverse the signs of terms involving β in (4.1.9)

$$L(u) = \begin{bmatrix} \gamma & 0 & 0 & -i\beta\gamma \\ 0 & 1 & 0 & 0 \\ 0 & 0 & 1 & 0 \\ i\beta\gamma & 0 & 0 & \gamma \end{bmatrix} \tag{4.4.3}$$

and then

$$p'_\lambda = \sum_\alpha L_{\lambda\alpha}(u)\, p_\alpha = \gamma m_0 u, 0, 0, \frac{i}{c}\gamma m_0 c^2$$

which is the expected result.

The manipulations given above represent an active transformation; in general they have conceptual advantages over the passive transformations and so these are the ones we shall use. The two types of transformation are displayed in figure 4.4 in terms of rotations in space-time. In diagram (a)

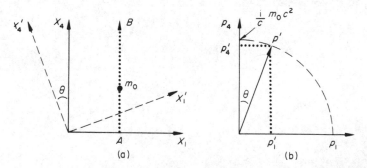

Figure 4.4

the dotted line represents the position of the body as a function of space and time when it is at rest in the Σ frame. The rotation through an angle θ establishes the Σ' frame moving with velocity $-u$ with respect to Σ (compare §2.6 and §3.3), and it is apparent that in the Σ' frame the path AB

represents a body moving with velocity

$$u = \frac{dx_1'}{dt'} = ic \frac{dx_1'}{dx_4'} = ic \tan\theta \qquad (4.4.4)$$

or

$$\tan\theta = -i\frac{u}{c}$$

in the $+X_1'$ direction. Figure 4.4 (b) represents the active transformation and in this case

$$\tan\theta = \frac{p_1'}{p_4'} = \frac{\gamma m_0 u}{(i/c)\, m_0 c^2} = -i\frac{u}{c} \qquad (4.4.5)$$

Thus $\theta = 0$ for bodies at rest.

An inspection of figure 4.4(b) shows that the momentum four-vector lies on the arc of a circle of radius $im_0 c$. This result follows immediately from the invariance property of four-vectors (4.1.5) and the Einstein equation (4.3.1)

$$\sum_{\lambda=1}^{4} p_\lambda^2 = \sum_{\lambda=1}^{4} p_\lambda'^2 = -m_0^2 c^2 \qquad (4.4.6)$$

The above equation is one for a four-dimensional sphere of radius $im_0 c$, and figure 4.4(b) represents the specialised case when the 2 and 3 (that is the y and z) components are zero. Despite the superficial unattractiveness of working with spheres with imaginary radii (or even nominally zero radii in the case of photons), this disadvantage is outweighed by the advantage of having a quick way of displaying the results of a Lorentz transformation— the reality of the physics in any case can always be found in the associated equations of the Lorentz transformation.

The equation (4.4.5)

$$\tan\theta = \frac{p_1'}{p_4'} = -i\frac{u}{c}$$

implies that θ has an upper limit of 45° when $u \to c$. If we now consider the situation in three dimensions we can include components p_x and p_y but must set $p_z = 0$ in every reference frame. (In practice we can use any two of the three components p_x, p_y, p_z.) The possible positions of the four-vector for a particle of rest mass m_0 are then represented by the shaded area of the sphere in figure 4.5 and Lorentz transformations merely move the radius vector around the surface of the allowed region. (A consideration of this figure and the nature of the radius $i\, m_0 c$ shows that the figure also effectively describes the behaviour of the four-velocity U_λ.) It is apparent that the limiting circumference (B), reached when $u \to c$, also represents the possible positions of the four-vector for massless particles.

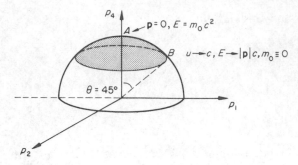

Figure 4.5 Allowed regions for the momentum four-vector (shaded). The letters A and B correspond to the physical limits displayed in figure 4.2.

4.5 Further examples of the Lorentz transformation of four-momenta

Let us firstly consider a Lorentz transformation in the X direction for a four-vector with components

$$p_\alpha = p_x, p_y, 0, i\frac{E}{c}$$

we then find, for a transformation corresponding to a velocity u, with the aid of (4.4.3)

$$p'_\lambda = \sum_\lambda L_{\lambda\alpha}(u)\, p_\alpha \qquad (4.5.1)$$

and

$$p'_x = p'_1 = \gamma\left(p_x + \beta\frac{E}{c}\right) \qquad (4.5.2)$$
$$p'_y = p'_2 = p_y$$
$$p'_z = p'_3 = 0$$
$$\frac{i\,E'}{c} = p'_4 = \frac{i}{c}\,\gamma(\beta p_x c + E)$$

where

$$\gamma = \frac{1}{\sqrt{(1 - u^2/c^2)}}, \qquad \beta = \frac{u}{c}$$

Next let us examine a transformation where the reference frames no longer possess relative motion along the X axis. This is the most general form of Lorentz transformation. Although we shall perform it for four-momenta, the method can be extended to any four-vector, including, of course, that for space-time. We shall examine a Lorentz transformation $L(\mathbf{u})$, where the direction of the vector $\mathbf{u}$ specifies the direction of the transformation. We again start with a four-vector p but instead of using components 1, 2, 3, 4 we write it as

$$p = \mathbf{p}_\| , \mathbf{p}_\perp , i\frac{E}{c} \tag{4.5.3}$$

where the terms $\mathbf{p}_\|$ and $\mathbf{p}_\perp$ denote the components of the space part, $\mathbf{p}$, of the four-vector parallel and perpendicular to $\mathbf{u}$ respectively.

With these components we use equations (4.5.2) again, in an appropriately modified notation

$$\mathbf{p}_\|' = \gamma \left(\mathbf{p}_\| + \boldsymbol{\beta}\frac{E}{c} \right)$$
$$\mathbf{p}_\perp' = \mathbf{p}_\perp \tag{4.5.4}$$
$$E' = \gamma (\boldsymbol{\beta}\cdot\mathbf{p}_\| c + E)$$

but since, with the notation $\boldsymbol{\beta}^2 = \beta^2$

$$\mathbf{p}_\| = \mathbf{p}\cdot\boldsymbol{\beta}\frac{\boldsymbol{\beta}}{\beta^2}$$

$$\mathbf{p}_\perp = \mathbf{p} - \mathbf{p}_\| = \mathbf{p} - \mathbf{p}\cdot\boldsymbol{\beta}\frac{\boldsymbol{\beta}}{\beta^2}$$

$$\mathbf{p}' = \mathbf{p}_\|' + \mathbf{p}_\perp'$$

then we can rewrite equation (4.5.4) as

$$\mathbf{p}' = \mathbf{p} + \boldsymbol{\beta}\left[(\gamma-1)\frac{\boldsymbol{\beta}\cdot\mathbf{p}}{\beta^2} + \gamma\frac{E}{c} \right] \tag{4.5.5}$$
$$E' = \gamma (\boldsymbol{\beta}\cdot\mathbf{p}c + E)$$

where the last equation arises from the fact that $\boldsymbol{\beta}\cdot\mathbf{p}_\perp = 0$ from the definition of $\mathbf{p}_\perp$.

The first line in equation (4.5.5) is often written in a slightly different form. Since

$$\beta^2 = \frac{\gamma^2 - 1}{\gamma^2} = \frac{(\gamma-1)(\gamma+1)}{\gamma^2}$$

then

$$\mathbf{p}' = \mathbf{p} + \boldsymbol{\beta}\left[\frac{\gamma^2}{\gamma+1}\boldsymbol{\beta}\cdot\mathbf{p} + \gamma\frac{E}{c} \right]$$

The arguments given above are not unique to the momentum four-vector, and so for any four-vector (4.1.6)

$$A = (\mathbf{A}, A_4)$$

we can adapt equations (4.5.5) to the form

$$\mathbf{A}' = \mathbf{A} + \boldsymbol{\beta} \left[\frac{\gamma^2}{\gamma+1} \boldsymbol{\beta} \cdot \mathbf{A} - i\gamma A_4 \right]$$
$$A_4' = \gamma (i\boldsymbol{\beta} \cdot \mathbf{A} + A_4)$$

(4.5.6)

The above equations allow us to assemble, without too great difficulty, the matrix of the Lorentz transformation (4.1.10)

$$A_\lambda' = \sum_{\alpha=1}^{4} L_{\lambda\alpha}(\mathbf{u})\, A_\alpha$$

in arbitrary directions. It is, with $\mathbf{u} = \boldsymbol{\beta}c$

$$L_{\lambda\alpha}(\mathbf{u}) = \begin{bmatrix} 1 + \beta_1^2 \dfrac{\gamma^2}{\gamma+1} & \beta_1\beta_2 \dfrac{\gamma^2}{\gamma+1} & \beta_1\beta_3 \dfrac{\gamma^2}{\gamma+1} & -i\beta_1\gamma \\[2ex] \beta_2\beta_1 \dfrac{\gamma^2}{\gamma+1} & 1 + \beta_2^2 \dfrac{\gamma^2}{\gamma+1} & \beta_2\beta_3 \dfrac{\gamma^2}{\gamma+1} & -i\beta_2\gamma \\[2ex] \beta_3\beta_1 \dfrac{\gamma^2}{\gamma+1} & \beta_3\beta_2 \dfrac{\gamma^2}{\gamma+1} & 1 + \beta_3^2 \dfrac{\gamma^2}{\gamma+1} & -i\beta_3\gamma \\[2ex] i\beta_1\gamma & i\beta_2\gamma & i\beta_3\gamma & \gamma \end{bmatrix}$$

(4.5.7)

It is left as an exercise to the reader to show that the above matrix satisfies equation (4.5.6). An inspection of the matrix (4.5.7) also shows that for transformations along the X direction, that is, when $\beta_1 = \beta$ and $\beta_2 = \beta_3 = 0$, then $L_{\lambda\alpha}$ reduces to the form given in (4.4.3).

Finally let us consider how to relate Lorentz transformations for the momentum four-vector in arbitrary directions to spherical geometry (compare figure 4.5). Consider first of all the two-dimensional plot given in figure 4.6 (a), where the radius AB is of length im_0c.

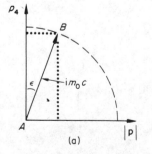 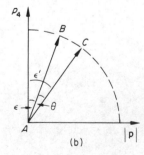

Figure 4.6

An inspection of this figure shows that

$$|\mathbf{p}| = im_0 c \sin \varepsilon$$

$$p_4 = im_0 c \cos \varepsilon \quad \text{or} \quad E = c \frac{p_4}{i} = m_0 c^2 \cos \varepsilon \qquad (4.5.8)$$

If the radius vector moves from B to C (figure 4.6(b)) then

$$|\mathbf{p}'| = im_0 c \sin \varepsilon' \qquad E' = m_0 c^2 \cos \varepsilon'$$

where

$$\varepsilon' = \varepsilon + \theta \qquad (4.5.9)$$

This situation is similar to that we have previously encountered for the transformation of velocities (3.7.6), and it is evident that θ is associated with the Lorentz boost $L(\mathbf{u})$ of the momentum four-vector. We can rewrite equation (4.5.9) as

$$\cos \varepsilon' = \frac{E'}{m_0 c^2} = \cos(\varepsilon + \theta)$$

$$= \cos \varepsilon \cos \theta - \sin \varepsilon \sin \theta \qquad (4.5.10)$$

$$= \frac{E}{m_0 c^2} \cos \theta - \frac{|\mathbf{p}|}{im_0 c} \sin \theta$$

We have seen previously (4.4.4) that the velocity associated with the Lorentz boost $L(\mathbf{u})$ can be written as

$$\tan \theta = -i \frac{u}{c} = -i\beta$$

hence (compare (2.6.13))

$$\cos \theta = \gamma, \qquad \sin \theta = -i\beta\gamma$$

and so equation (4.5.10) becomes

$$\cos \varepsilon' = \frac{E'}{m_0 c^2} = \frac{E\gamma}{m_0 c^2} + \frac{|\mathbf{p}|}{im_0 c} i\beta\gamma$$

or

$$E' = \gamma(\beta |\mathbf{p}| c + E) \qquad (4.5.11)$$

This equation is equivalent to the second of equations (4.5.5) when $\boldsymbol{\beta}$ and $\mathbf{p}$ are parallel. Next let us examine a more complicated problem. Consider the two Lorentz transformations displayed in figure 4.7. We start with a body at rest at A with rest mass m_0, and the initial Lorentz transformation takes the momentum four-vector to B, where the arc AB subtended at O defines the angle ε appearing in equations (4.5.8) and (4.5.10). A second Lorentz transformation takes the momentum four-vector from B to C so that the angle ε'

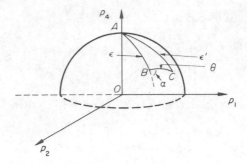

Figure 4.7

associated with AC defines the final four-momentum of the body. The angle associated with the arc BC we shall label as θ, and the angle between the directions AB and BC we call α.

The properties of spherical triangles (appendix 3) yield the equation

$$\cos\varepsilon' = \cos\varepsilon\,\cos\theta - \sin\varepsilon\,\sin\theta\,\cos\alpha \qquad (4.5.12)$$

and it is immediately apparent that for two successive transformations in the direction AB, so that $\alpha = 0$, then

$$\varepsilon' = \varepsilon + \theta$$

This is the situation we have already encountered above in equations (4.5.9) to (4.5.11). For $\alpha \neq 0$ we may use equations (4.5.10) and (4.5.11) to rewrite (4.5.12) as

$$\frac{E'}{m_0 c^2} = \frac{E}{m_0 c^2}\cdot\gamma - \frac{p}{im_0 c}\cdot -i\beta\gamma\cdot\cos\alpha$$

and if we identify α with the angle between $\boldsymbol{\beta}$ and $\mathbf{p}$

$$E' = \gamma(E + \beta c\mathbf{p}\cdot\boldsymbol{\beta})$$

and we see that we have recovered the second of equations (4.5.5). The first of the equations (4.5.5) can also be established by considering projections on the base of the hemisphere (but the process is laborious). An alternative, and simpler, approach is given in appendix 3.

An examination of figure 4.7 also reveals another important feature concerning successive Lorentz transformations, namely that one must define the order in which the transformations are carried out since transformations in reverse order do not take one to the same point on the sphere. We illustrate this point in figure 4.8 where for convenience of display we have tilted the diagram and also chosen the angle α to be 90°. The initial transformation is ABC. The reverse transformation is $AB'C'$—thus the same point is not

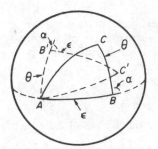

Figure 4.8

reached. Only if both transformations are in the same direction (§3.8) is the same final point reached. The property that

$$L_1 L_2 \neq L_2 L_1 \qquad (4.5.13)$$

where L_1 and L_2 are Lorentz transformations is normally described by saying that successive Lorentz transformation do not *commute*.*

4.6   The emission of light from moving objects †

Let us assume that we have a source of light of frequency v and that an observer in the same reference frame as the light says that it is emitted isotropically—that is, with equal probability in all directions. We shall designate this reference frame as Σ (figure 4.9(a)), and ask what a second

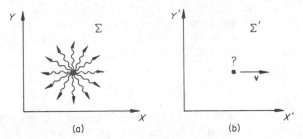

Figure 4.9

observer sees in a reference frame Σ' (figure 4.9(b)) in which the source moves

---

* Operations which commute have the property
$$A_1 A_2 - A_2 A_1 = 0$$
where the A's denote mathematical operators.
† Much of the material discussed in this section (§4.6.1 and 4.6.2) can also be treated by starting with a plane wave with amplitude
$$\psi \sim \exp[i(\mathbf{k} \cdot \mathbf{r} - \omega t)]$$
and exploiting the Lorentz invariance of the phase (this treatment is common in textbooks on special relativity). However the intensity of the emitted light (§4.6.3) cannot easily be handled in this manner. It is of interest to note that a treatment based on mechanical concepts leads to exactly the same conclusions as the wave picture of light for aberration and the Doppler shift.

with a velocity **v** in the X direction. As we stated in §4.4, the net result can be achieved by a passive transformation of coordinates equivalent to $-\mathbf{v}$ or by an active transformation of momentum four-vectors. We shall use the latter method, and in this case the momentum four-vectors are those for the outgoing photons.

4.6.1   The aberration of light

Let us concentrate on the behaviour of a single photon. We shall represent it by the four-vector*

$$k = (\mathbf{k}, i\omega/c)$$

in the rest frame (Σ) of the light source (figure 4.10(a)). In this frame let the photon be emitted at an angle ψ to the X axis. In the coordinate frame of the

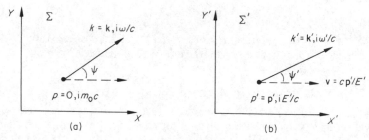

Figure 4.10

observer (Σ'), the same quantities are denoted by primes (figure 4.10(b)). We shall assume that the source of light is body with rest mass m_0, and shall represent it by the four-vectors p and p' in Σ and Σ' respectively.

An inspection of figure 4.10 shows that the angles of emission are determined by the relations

$$\tan\psi = k_y/k_x \qquad \tan\psi' = k'_y/k'_x$$

Now we may adapt equations (4.5.2) to the present situation by the substitution of $\mathbf{k}$ and ω for $\mathbf{p}$ and E respectively; we then find

$$\tan\psi' = \frac{k'_y}{k'_x} = \frac{k_y}{\gamma(k_x + \beta\omega/c)} \tag{4.6.1}$$

where βc is the velocity of the source. If we write

$$k_x = |\mathbf{k}|\cos\psi \qquad k_y = |\mathbf{k}|\sin\psi$$

and use the fact that since the photon is massless the Einstein equation (4.3.2) becomes

$$\omega = |\mathbf{k}|\,c$$

---

* Here **k** is the linear momentum of the photon, and ω is its energy (compare (4.2.7) and (4.2.8)).

then equation (4.6.1) reduces to

$$\tan\psi' = \frac{\sin\psi}{\gamma\,(\cos\psi + \beta)} \tag{4.6.2}$$

This is the relativistic formula for the aberration of light; it was first deduced by Einstein in 1905. The phenomenon of the aberration of light was known long before 1905, however, albeit in a nonrelativistic context. Consider figure 4.11 in which light is emitted at an angle ψ in the Σ frame and received

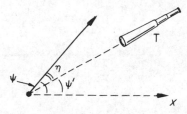

Figure 4.11

by a telescope T tilted an angle ψ' in the Σ' frame. Let the difference between ψ' and ψ be η, and for reasons which will appear below we shall assume that η is very small. In the nonrelativistic limit $\beta \to 0$, $\gamma \to 1$, and equation (4.6.2) can then be written as

$$\tan(\psi - \eta) = \tan\psi - \eta = \frac{\sin\psi}{\cos\psi\,(1 + \beta/\cos\psi)}$$
$$\xrightarrow[\beta \to 0]{} \tan\psi\,(1 - \beta/\cos\psi)$$

thus

$$\eta = \beta\,\sin\psi \tag{4.6.3}$$

For $\beta \to 0$ this justifies our use of the approximation of very small η in the above derivation.

The aberration of light was first observed by Bradley,* who made systematic observations on the star γ-Draconis for a period of one year. During this time he found that the tilt of his telescope had to be changed in a regular fashion. This arose because of the earth's motion about the sun which we illustrate schematically in figure 4.12. During the course of the year the earth's velocity with respect to the direction of γ-Draconis changes between the two limits indicated by the positions A and B. Consequently one expects an annual wobble in the tilt of the telescope since the limits on β change from $\beta = -v/c$ to $\beta = +v/c$ where v is the velocity of the earth around the sun and c is the velocity of light. Bradley correctly deduced the cause of the wobble·and calculated that its magnitude could be explained if light took 8 min 13 s to travel between the earth and sun with an error of 5 to 10 s (that is, roughly

---

* *Phil. Trans. Roy. Soc.* **35**, 637, 1728.

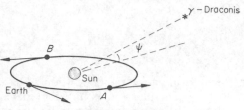

Figure 4.12

1 to 2 per cent accuracy). This represents a value of 3×10^8 km/s for c and was by far the most accurate determination of c at that time. A facsimile of his results are displayed in table 4.2

1727.	D.	The Difference of Declination by Obſervation. "	The Difference of Declination by the Hypotheſis. "	1728.	D.	The Difference of Declination by Obſervation. "	The Difference of Declination by the Hypotheſis. "
October 20th --		4½	4¼	March •	24	37	38
November -	17	11½	12	April - -	6	36	36½
December -	6	17½	18½	May - -	6	28½	29½
- - -	28	25	26	June - -	5	18½	20
1728				- - -	15	17½	17
January -	24	34	34	July - -	3	11½	11½
February -	10	38	37	Auguſt -	2	4	4
March - -	7	39	39	September -	6	0	0

Table 4.2

4.6.2 The Doppler Shift

Next let us examine how the frequency of the light changes during emission from the moving source. Let the rest mass of the source by m_0, and let it be represented by the four-vector p_λ with space and fourth components 0 and im_0c respectively (figure 4.11). Similarly for the photon we use a four-vector k_λ with space and fourth components $\mathbf{k}$ and $i\omega/c$ respectively. Now the scalar product $\sum_\lambda p_\lambda k_\lambda$ is Lorentz invariant (4.1.5), and, if we represent the rest frame of the source by Σ, we can write

$$\sum_{\lambda=1}^{4} p_\lambda k_\lambda = -\frac{1}{c^2} m_0 c^2 \omega \qquad \text{in the } \Sigma \text{ frame}$$

$$= \mathbf{p}' \cdot \mathbf{k}' - \frac{1}{c^2} E' \omega' \qquad \text{in the } \Sigma' \text{ frame} \qquad (4.6.4)$$

$$= \gamma m_0 \beta c |\mathbf{k}'| \cos \psi' - \frac{\gamma}{c^2} m_0 c^2 \omega'$$

since $\mathbf{p}=0$ in the rest frame. For photons, the Planck and Einstein relations

$$\omega = hv \qquad |\mathbf{k}| = \frac{\omega}{c} = \frac{hv}{c}$$

where h is Planck's constant and v the frequency of light associated with the photon, imply that we can rewrite equation (4.6.4) as

$$m_0 hv = \gamma m_0 hv' - \gamma m_0 \beta hv' \cos \psi'$$

Hence

$$v = v'\gamma (1 - \beta \cos \psi')$$

or

$$v' = \frac{v}{\gamma (1 - \beta \cos \psi')} \tag{4.6.5}$$

This is the equation of the relativistic Doppler effect. An inspection of the equation shows that as the body moves towards the observer and then away from him the frequency of the detected light will steadily fall in value as ψ' changes between 0 and π. Thus the light moves from the blue to the red end of the spectrum. This principle ('the red shift') was used by the astronomer Hubble to measure the velocity of receding galaxies in the universe and thus to establish Hubble's law—this is a linear relationship between the velocity of the receding galaxy and its apparent distance from the earth. An interesting feature of equation (4.6.5) is that the final result is independent of the rest mass of the light source, and of its direction of emission (ψ) in the Σ frame.

4.6.3   The intensity of the emitted light

Lastly let us consider how the intensity of the light changes in going between the Σ and Σ' frames. We postulated originally that the light was emitted isotropically by the body at rest in the Σ frame. Thus if we consider figure 4.13

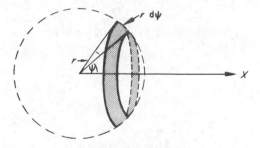

Figure 4.13

and assume that N photons are emitted in all directions, then the number travelling in the direction $\psi \rightarrow \psi + d\psi$ (the shaded ring in figure 4.11) is

$$dN = \frac{N \, 2\pi r \sin \psi \, r \, d\psi}{4\pi r^2}$$

where r is the radius of any arbitrary sphere drawn around the source. The above equation merely says that the probability of the photon's passing through any element of the surface area is the same in all directions if the emission is isotropic. We may rewrite the equation as

$$\frac{dN}{2\pi \sin\psi \, d\psi} = \frac{dN}{d\Omega} = \frac{N}{4\pi} \qquad (4.6.6)$$

where $d\Omega$ is called the element of solid angle

$$d\Omega = 2\pi \sin\psi \, d\psi = -2\pi \, d(\cos\psi) \qquad (4.6.7)$$

Since the surface of the sphere subtends a solid angle of 4π steradians at the centre, $N/4\pi$ represents the number of photons per unit solid angle.

Now if we transfer our attention to the Σ' frame, the dN photons will be emitted in the angular interval $\psi' \to \psi' + d\psi'$ and so that number per unit solid angle becomes $dN/d\Omega'$, where by (4.6.7)

$$\frac{d\Omega}{d\Omega'} = \frac{d(\cos\psi)}{d(\cos\psi')} \qquad (4.6.8)$$

We must now seek a relation between $\cos\psi$ and $\cos\psi'$. If we recall equation (4.1.12)

$$A_\lambda = \sum_\alpha L_{\alpha\lambda} A'_\alpha$$

and apply it to the momentum four-vectors k and k' with the aid of the transformation matrix (4.4.3), then we find

$$
\begin{aligned}
k_1 &= \gamma(k'_1 + i\beta k'_4) \\
k_4 &= \gamma(-i\beta k'_1 + k'_4)
\end{aligned}
\quad \text{or} \quad
\begin{aligned}
|\mathbf{k}| \cos\psi &= \gamma |\mathbf{k}'| (\cos\psi' - \beta) \\
|\mathbf{k}|c &= \gamma |\mathbf{k}'|c(-\beta \cos\psi' + 1)
\end{aligned}
$$

hence if we divide one equation by the other, we obtain

$$\cos\psi = \frac{\cos\psi' - \beta}{1 - \beta \cos\psi'} \qquad (4.6.9)$$

and so

$$\frac{d(\cos\psi)}{d(\cos\psi')} = \frac{1 - \beta^2}{(1 - \beta \cos\psi')^2}$$

Thus the number of photons emitted per unit solid angle between ψ' and $\psi' + d\psi'$ is

$$\frac{dN}{d\Omega'} = \frac{dN}{d\Omega} \frac{d\Omega}{d\Omega'} = \frac{N(1 - \beta^2)}{4\pi(1 - \beta \cos\psi')^2} \qquad (4.6.10)$$

Thus as $\beta \to 1$ the light becomes more and more concentrated in the forward direction. If we combine the information in (4.6.10) with that for the Doppler effect (4.6.5), we see that as bodies approach an observer at velocities tending to those of light they appear 'bright and blue' and as they pass him and recede they become 'dim and red'.

4.7 The addition of four-vectors

In three-dimensional space the addition of two vectors yields a third. For example if we have vectors $\mathbf{p}_a$ and $\mathbf{p}_b$, representing the momenta of two particles (figure 4.14), then their resultant is given by

$$\mathbf{P} = \mathbf{p}_a + \mathbf{p}_b \tag{4.7.1}$$

where the x, y and z components of $\mathbf{p}_a$ and $\mathbf{p}_b$ add (for convenience we assume

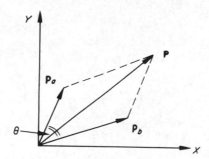

Figure 4.14

$p_{az} = p_{bz} = 0$)

$$P_x = p_{ax} + p_{bx}$$
$$P_y = p_{ay} + p_{by}$$
$$P_z = p_{az} + p_{bz} = 0$$

also (compare (4.1.1).)

$$\mathbf{P} = \mathbf{i}P_x + \mathbf{j}P_y + \mathbf{k}P_z$$

The length of $\mathbf{P}$ is given by the cosine formula

$$\mathbf{P}^2 = (\mathbf{p}_a + \mathbf{p}_b)^2 = \mathbf{p}_a^2 + 2\mathbf{p}_a \cdot \mathbf{p}_b + \mathbf{p}_b^2$$
$$= \mathbf{p}_a^2 + 2|\mathbf{p}_a|\,|\mathbf{p}_b|\cos\theta + \mathbf{p}_b^2 \tag{4.7.2}$$

where θ is the angle between $\mathbf{p}_a$ and $\mathbf{p}_b$ (figure 4.14).

If the particles a and b collide, conservation of momentum gives

$$\mathbf{p}_a + \mathbf{p}_b = \mathbf{P} = \mathbf{p}'_a + \mathbf{p}'_b \tag{4.7.3}$$

where the dashes denote the final states. At the same time, conservation of energy gives

$$E_a + E_b = E = E'_a + E'_b \tag{4.7.4}$$

Equations (4.7.3) and (4.7.4) appear in Newtonian mechanics, but they can be applied equally well in relativistic mechanics, providing we interpret **P** and E as components of four-vectors. We therefore represent (4.7.3) and (4.7.4) by the single equation

$$p_{a\lambda} + p_{b\lambda} = P_\lambda = p'_{a\lambda} + p'_{b\lambda} \tag{4.7.5}$$

and since P_λ is a four-vector then P_λ^2 is an invariant quantity ('a Lorentz scalar') in all reference frames. The Lorentz invariance property of P_λ^2 is extremely useful in elementary particle physics.

4.7.1 Compton scattering

Let us apply equation (4.7.5) to the scattering of photons by electrons – this is known as the Compton effect. We shall denote the four-vectors for the photon by the symbol k and those for the electron by p. Then equation (4.7.5) becomes

$$k_\lambda + p_\lambda = k'_\lambda + p'_\lambda \tag{4.7.6}$$

We may rewrite this equation as

$$\sum_{\lambda=1}^{4} (k_\lambda + p_\lambda - k'_\lambda)^2 = \sum_{\lambda=1}^{4} p'^2_\lambda = -m_e^2 c^2 \tag{4.7.7}$$

where m_e is the rest mass of the electron (compare (4.3.1)). We evaluate the left-hand side of the above equation and find

$$-m_e^2 c^2 + \sum_\lambda (2k_\lambda p_\lambda - 2k_\lambda k'_\lambda - 2p_\lambda k'_\lambda) = -m_e^2 c^2$$

or

$$\sum_\lambda k_\lambda k'_\lambda = \sum_\lambda k_\lambda p_\lambda - \sum_\lambda p_\lambda k'_\lambda \tag{4.7.8}$$

The individual four-vectors in the above equation can be written in the manner of equation (4.1.6)

$$\begin{array}{ll} \text{incident photon } k = (\mathbf{k}, i\omega/c) & \\ \text{scattered photon } k' = (\mathbf{k}', i\omega'/c) & (4.7.9) \\ \text{incident electron } p = (\mathbf{p}, iE/c) & \end{array}$$

and since the photons are massless the Einstein equation (4.3.2) also gives the relations

$$\omega = |\mathbf{k}|\, c \qquad \omega' = |\mathbf{k}'|\, c$$

If we denote the angle between the vectors **k** and **k′** by θ (the angle of scattering) the individual terms in (4.7.8) become

$$\sum_\lambda k_\lambda k'_\lambda = \mathbf{k}\cdot\mathbf{k}' - \frac{\omega\omega'}{c^2} = \frac{\omega\omega'}{c^2}(\cos\theta - 1)$$

$$\sum_\lambda k_\lambda p_\lambda = \mathbf{k}\cdot\mathbf{p} - \frac{\omega E}{c^2} \qquad (4.7.10)$$

$$\sum_\lambda p_\lambda k'_\lambda = \mathbf{p}\cdot\mathbf{k}' - \frac{E\omega'}{c^2}$$

So far we have made no mention of possible reference frames in which to examine the process. We shall work in the laboratory reference frame where the electron is initially at rest, then

$$\mathbf{p}=0 \qquad E=m_ec^2$$

and combining equations (4.7.8) and (4.7.10) yields

$$\frac{\omega\omega'}{c^2}(\cos\theta - 1) = -\frac{m_ec^2}{c^2}(\omega - \omega') \qquad (4.7.11)$$

Now the Planck relation between the energy of photons and their frequency, v, and wavelength, λ, is

$$\omega = hv = h\frac{c}{\lambda} \qquad (4.7.12)$$

where h is Planck's constant. If we divide equation (4.7.11) throughout by $\omega\omega'$ it is not difficult to show that

$$\lambda' = \lambda + \frac{h}{m_ec}(1 - \cos\theta)$$

$$= \lambda + \frac{2h}{m_ec}\sin^2\frac{\theta}{2} \qquad (4.7.13)$$

Thus it can be seen that if light is scattered from electrons, the wavelength of the scattered radiation increases with the scattering angle.

This result is known as the Compton effect, after its discoverer A. H. Compton,* who both developed the theory and obtained the experimental data. A comparison of calculation and experiment from Compton's paper of 1923 is shown in figure 4.15; γ-rays from RaC were used for the incident photon beam. The good agreement between experiment and theory was confirmed in subsequent and more refined experiments.

---

* COMPTON, A. H., *Phys. Rev.* **21**, 483, 1923; *ibid.* **22**, 409, 1923 and *Phil. Mag.* **41**, 760, 1923.

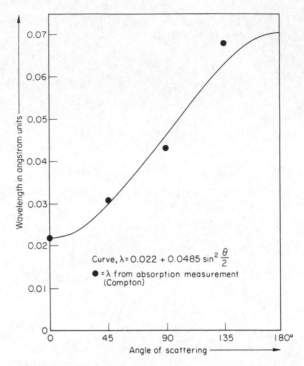

Figure 4.15 Wavelength of scattered RaC γ-rays as a function of angle in Compton's original experiment

4.7.2 Elastic antiproton-proton scattering

The bubble chamber technique, which permits the photography of the tracks of charged particles in liquids, offers many nice examples of the kinematics of nuclear particles when they interact with each other.

In this section we shall examine the opening angle between a proton and an antiproton* after an elastic collision in which an incoming antiproton has struck a proton in a hydrogen bubble chamber (the struck proton is a nucleus of one of the hydrogen atoms in the bubble chamber liquid). Examples of these collisions at different antiproton momenta are shown in figure 4.16, and display a steady collapse of the size of the opening angle with increasing antiproton momenta. This collapse arises solely from the relativistic factors, and so the pictures provide a visual demonstration of the correctness of Einstein's work.

The equation of conservation of momentum is from (4.7.3)

$$\mathbf{p}_a + \mathbf{p}_b = \mathbf{p}'_a + \mathbf{p}'_b$$

We shall designate the antiproton by a and the proton by b. Since the proton

---

* The proton has charge e and Mass $M = 938$ MeV/c^2; the antiproton has charge $-e$ and an identical mass.

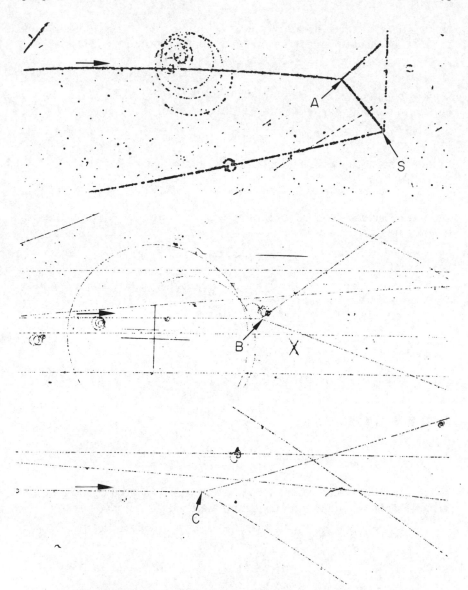

Figure 4.16 Collisions of antiprotons with protons at rest, (A) an antiproton with near zero momentum, (B) an antiproton of incident momentum 4.5 GeV/c and (C) an antiproton with momentum 9.0 GeV/c. The expected opening angles are 90° (A), 60° (B) and 46° (C). The directions of the incoming antiprotons are marked by arrows. In the collision (A), the antiproton subsequently annihilates with another proton at the point S to yield mesons

is stationary in the laboratory reference frame before the collision, then $\mathbf{p}_b = 0$, and we can rewrite the above equation in the notation of (4.7.2)

$$\mathbf{p}_a^2 = \mathbf{p}_a'^2 + 2\mathbf{p}_a' \cdot \mathbf{p}_b' + \mathbf{p}_b'^2 \qquad (4.7.14)$$

Now the interactions displayed in figure 4.16 have been selected so that the moduli of the outgoing momenta are roughly equal

$$|\mathbf{p}_a'| = |\mathbf{p}_b'| = p$$

In these circumstances equation (4.7.14) becomes

$$\mathbf{p}_a^2 = 2p^2 (1 + \cos \theta) = 4p^2 \cos^2 \theta/2 \qquad (4.7.15)$$

and if we designate the mass of proton and antiproton by M, the energies of the outgoing particles are given by

$$E_a'^2 = E_b'^2 = c^2 (p^2 + M^2c^2) = \tfrac{1}{4}(E_a + Mc^2)^2 \qquad (4.7.16)$$

where we have made use of the Einstein relation (4.3.2), and also the equation of the conservation of energy (4.7.4)

$$E_a + E_b = E_a + Mc^2 = E_a' + E_b' = 2E_a'$$

Equation (4.7.16) may be solved for p^2

$$\begin{aligned} 4p^2 &= (1/c^2) (E_a^2 + 2E_aMc^2 - 3M^2c^4) \\ &= [\mathbf{p}_a^2 + 2M (E_a - Mc^2)] \end{aligned}$$

hence from (4.7.15)

$$\cos^2 \theta/2 = \frac{\mathbf{p}_a^2}{\mathbf{p}_a^2 + 2M (E_a - Mc^2)} \qquad (4.7.17)$$

Let us first consider the nonrelativistic limit $\mathbf{p}_a \to 0$, then, using the binomial expansion the second term of the denominator in (4.7.17) involves

$$\begin{aligned} E_a - Mc^2 &= \frac{Mc^2}{\sqrt{(1 - u^2/c^2)}} - Mc^2 \\ &= Mc^2 (1 + \tfrac{1}{2}u^2/c^2) - Mc^2 \\ u/c &\to 0 \end{aligned}$$

and so

$$2M (E_a - Mc^2) = 2M\tfrac{1}{2}Mu^2 = \mathbf{p}_a^2$$
$$u/c \to 0$$

hence from (4.7.17)

$$\cos^2 \theta/2 \to \tfrac{1}{2} \quad \text{for} \quad u/c \to 0$$

Thus the half angle $\theta/2$ is $45°$ and the opening angle $\theta = 90°$. This is the result found in the familiar problem of the collision of billiard balls in classical non-relativistic mechanics. Its equivalent in elementary particle physics is shown in figure 4.16 at the point A for an antiproton with near zero momentum. The opening angle is $\sim 90°$; in contrast the vertices B and C refer respectively to events involving antiprotons with momenta of 4.5 and 9.0 GeV/c respectively and the opening angles are $\sim 60°$ and $46°$. The photographs in figure 4.16 show two-dimensional projections of events which took place in three dimensions – the bubble chamber is viewed by three separate cameras and the event in space is reconstructed from the two-dimensional pictures. It is not possible to select events which are completely perpendicular to a camera axis, and so a single picture can only give a good approximation to the true kinematic angle. All three photographs were taken at the CERN laboratory in Geneva and were analysed at the University of Liverpool. In performing the kinematic reconstruction of the events, determinations are made of both the angles of the outgoing particles with respect to the incident antiproton and also the momenta of all three tracks. The latter are determined from their curvature in the magnetic field which is applied across the chamber (compare §6.5.2).

4.8   The centre of momentum system

In the previous section we examined the Compton effect for an electron at rest in the laboratory system. We shall now examine the process again for the situation where the electron is also in motion.

The conservation of energy and momentum in the collision was expressed in a Lorentz invariant form in equation (4.7.8)

$$\sum_\lambda k_\lambda k'_\lambda = \sum_\lambda k_\lambda p_\lambda - \sum_\lambda p_\lambda k'_\lambda$$

This equation may be rewritten as

$$\sum_\lambda k'_\lambda (k+p)_\lambda = \sum_\lambda k_\lambda p_\lambda \qquad (4.8.1)$$

and if we use the Einstein relation (4.3.2) to express $\mathbf{k}'$ (the photon momentum) as

$$\mathbf{k}' = \frac{\omega'}{c} \mathbf{f}$$

where $\mathbf{f}$ is a unit vector pointing along the direction of $\mathbf{k}'$, then, with the aid of (4.7.9), equation (4.8.1) becomes

$$\frac{\omega'}{c}\left[\frac{\omega + E}{c} - (\mathbf{k}+\mathbf{p})\cdot\mathbf{f}\right] = -\sum_\lambda k_\lambda p_\lambda \qquad (4.8.2)$$

Now consider the term $(k+p)_\lambda$ in equation (4.8.1). In equation (4.8.2) we have expressed this four-vector in a form suitable for the laboratory frame.

However, another very useful reference frame exists in atomic and nuclear physics – the *centre of momentum frame* in which the net linear momentum of the incident particles is zero; that is, in the present circumstances

$$\mathbf{k}_c + \mathbf{p}_c = 0 \qquad \text{or} \qquad \mathbf{k}_c = -\mathbf{p}_c \qquad (4.8.3)$$

where we have used the subscript c to indicate momenta in the centre of momentum frame (henceforward it will be called the *c-frame* or *c-system*). Thus the four-vector $(k+p)$ assumes the form (compare (4.1.6))

$$k+p = \mathbf{k}+\mathbf{p}, (i/c)\,(\omega+E) \text{ in the laboratory system}$$
$$= 0, (i/c)\,(\omega_c + E_c) \text{ in } c\text{-system} \qquad (4.8.4)$$

A comparison with the basic equations for the momentum four-vector ((4.2.8) and (4.2.10)) and the Einstein equation (4.3.2) then shows that $(\omega_c + E_c)/c^2$ is the effective 'rest mass' associated with the four-vector $(k+p)$ (compare (4.4.1)), and that this mass moves with a velocity u in the laboratory frame where

$$\frac{\mathbf{u}}{c} = \boldsymbol{\beta} = \frac{c\,(\mathbf{k}+\mathbf{p})}{\omega+E} \qquad (4.8.5)$$

The Lorentz invariance of the square of the four-vector $(k+p)$ also gives, from (4.8.4)

$$\sum_{\lambda=1}^{4} (k+p)_\lambda^2 = (\mathbf{k}+\mathbf{p})^2 - (1/c^2)\,(\omega+E)^2$$
$$= 0 - (1/c^2)\,(\omega_c^2 + E_c^2) \qquad (4.8.6)$$

or

$$\omega_c + E_c = (\omega+E)\sqrt{\left[1 - \frac{c^2\,(\mathbf{k}+\mathbf{p})^2}{(\omega+E)^2}\right]}$$
$$= (\omega+E)\sqrt{(1-\beta^2)}$$
$$= (\omega+E)/\gamma \qquad (4.8.7)$$

Let us now return to equation (4.8.2); we find from equations (4.8.5) and (4.8.7) that (4.8.2) becomes

$$\frac{\omega'}{c^2}\,\gamma(\omega_c + E_c)\,(1-\boldsymbol{\beta}\cdot\mathbf{f}) = -\sum_\lambda k_\lambda p_\lambda \qquad (4.8.8)$$

Equation (4.8.3) implies that the right-hand side can be written in the c-system as

$$-\sum_\lambda k_\lambda p_\lambda = \frac{\omega_c E_c}{c^2} - \mathbf{k}_c \cdot \mathbf{p}_c = \frac{\omega_c E_c}{c^2} + \mathbf{k}_c^2$$
$$= \frac{\omega_c}{c^2}\,(E_c + \omega_c)$$

Practical situations involving forces in special relativity normally appear in electromagnetism and so we shall postpone further discussion of them until chapter 6. We close this chapter by mentioning four-tensors. Within the framework of the general Lorentz transformation a four-vector is a single member of a system of four-tensors which transform as

$$T'_{\lambda\mu\varrho\ldots} = \sum_{\alpha\beta\gamma\ldots} L_{\lambda\alpha}L_{\mu\beta}L_{\varrho\gamma}T_{\alpha\beta\gamma\ldots} \tag{4.9.5}$$

where the indices run from 1 to 4 and the components of the matrix L are the elements of the Lorentz transformation (compare (4.1.10)).

A tensor of rank zero has one component and remains invariant (that is, scalar) under Lorentz transformations—an example we have already encountered is the scalar product of two four-vectors

$$T_0 \equiv \sum_\lambda T'_{\lambda\lambda} = \sum_\lambda A'_\lambda B'_\lambda = \sum_\alpha T_{\alpha\alpha} \tag{4.9.6}$$

A four-tensor of rank one has four components which transform as

$$T'_\lambda = \sum_{\alpha=1}^{4} L_{\lambda\alpha}T_\alpha \tag{4.9.7}$$

and is thus a four-vector with a different name. A four-tensor of rank two transforms as

$$T'_{\lambda\mu} = \sum_{\alpha\beta} L_{\lambda\alpha}L_{\mu\beta}T_{\alpha\beta} \tag{4.9.8}$$

In the next chapter we shall encounter tensors of rank two when discussing angular momentum. A further example is the derivative of a four vector A_λ

$$\frac{\partial A_\lambda}{\partial x_\mu} = T_{\lambda\mu}$$

since, by (4.1.12) and (4.9.7)

$$\begin{aligned}
T'_{\lambda\mu} &= \frac{\partial A'_\lambda}{\partial x'_\mu} = \frac{\partial x_\beta}{\partial x'_\mu}\frac{\partial A'_\lambda}{\partial x_\beta} \\
&= \sum_{\beta\alpha} L_{\mu\beta}L_{\lambda\alpha}\frac{\partial A_\alpha}{\partial x_\beta} \\
&= \sum_{\alpha\beta} L_{\lambda\alpha}L_{\mu\beta}T_{\alpha\beta}
\end{aligned} \tag{4.9.9}$$

The tensor $T_{\alpha\beta}$ is said to be symmetric if $T_{\alpha\beta} = T_{\beta\alpha}$ and antisymmetric if $T_{\alpha\beta} = -T_{\beta\alpha}$. Thus the diagonal components of an antisymmetric tensor must vanish, since the condition $T_{\alpha\alpha} = -T_{\alpha\alpha}$ can only be fulfilled if $T_{\alpha\alpha} = 0$.

Tensors of higher rank than two do not normally appear in the theory of special relativity and so the subject will not be pursued further.

PROBLEMS

4.1 Use the Lorentz transformation formula (4.1.2) to demonstrate the Lorentz invariance of the scalar product of four-vectors (4.1.5).

4.2 One atomic mass unit (1 a.m.u.) is equal to 1.66×10^{-27} kg. The rest masses of the proton and neutron are 1.00731 and 1.00867 a.m.u. respectively. These particles combine to give a deuteron which has mass 2.01360 a.m.u. This figure is smaller than the combined masses of the neutron and proton because part of their mass disappears as binding energy in the combined system. What is the binding energy in (a) joule (b) MeV? (1 MeV = 1.6×10^{-13} joule).

4.3 A particle in a certain reference frame has total energy $0.938 \sqrt{2}$ GeV and momentum 0.938 GeV/c. What is its rest mass in GeV/c^2 and a.m.u.? What is its total energy in a second frame where its momentum is 0.5 GeV/c, and what is the relative velocity of the two reference frames? The appropriate conversion units are given in question 4.2.

4.4 A particle X at rest decays to two particles a and b. Conservation of momentum implies that the two bodies fly off with equal and opposite momentum $q = p_c$ in the rest frame of X. Show that

$$q = \frac{c}{2M_x} \sqrt{\{[M_x^2 - (m_a + m_b)^2][M_x^2 - (m_a - m_b)^2]\}}$$

where M_x, m_a and m_b are the rest masses of X, a and b respectively.

4.5 In the previous question show that the momentum of a in the rest frame of b is $q M_x/m_b$. (*Hint*: First find a Lorentz transformation which brings b to rest).

4.6 The lifetime of the π meson in its own rest frame is 2.6×10^{-8} s. What fraction of a beam of pions would survive a journey 200 m along a vacuum pipe at a momentum of 1.4 GeV/c in (a) the absence (b) the presence of the time dilatation factor? The rest mass energy of the pion is 0.14 GeV/c^2.

4.7 Consider the following collision of elementary particles

$$ab \to de$$

where b is at rest in the laboratory frame.

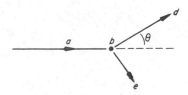

What is the direction of a in the rest frame of d?

4.8 In the CERN I.S.R. (intersecting storage ring) two beams of proton each with momentum 25 GeV/c collide at an angle of $\sim 15°$.

What is the energy of the protons in beam b in the rest frame of protons of beam a? As a good approximation, set the mass of a proton to be 1 GeV/c^2.

4.9 A particle with four-momentum components $p_x, p_y, 0, iE/c$ is given a Lorentz boost corresponding to β_1 in the X direction followed by one of β_2 in the Y direction. A second particle of equal four-momentum receives sequential boosts corresponding to β_2 and β_1 in the X and Y directions respectively. What are the final energies resulting from the two operations?

4.10 A proton, p, strikes a second proton at rest in the laboratory system and produces an antiproton, $\bar{\text{p}}$, in the reaction

$$\text{pp} \rightarrow \text{ppp}\bar{\text{p}}$$

What is the threshold momentum in the laboratory frame for this reaction to occur? You may take the mass of proton and antiproton to be 1 GeV/c^2.

4.11 How fast must a body be travelling for red light (wavelength 7000 A.U.) to appear blue (4000 A.U.) to an observer in its approach path?

4.12 A small cube, A, (of sides 1 mm with its front face perpendicular to its line of flight) has a velocity of $0.999c$ towards a similar cube B. In the rest frame of B an observer O is placed 1 m from the cube B. (see figure).

At a certain time the observer first records seeing the back face of cube A. Where was cube A, measured in the frame of O, when it emitted the light that the observer interpreted as this event? After seeing the back of cube A can the observer warn cube B that it is going to be hit by cube B before it is actually hit?

4.13 Show that equation (4.2.5), $\sum_\lambda U_\lambda^2 = -c^2$, implies that $\sum_\lambda U_\lambda A_\lambda = 0$ in every reference frame, where A_λ is the four-acceleration. Hence deduce that $\sum_\lambda A_\lambda^2 = \mathbf{a}^2$ where $\mathbf{a}$ is the acceleration in the rest frame of the body.

4.14 A rocket leaves the earth at time $t=0$, and travels in a straight line with constant acceleration a relative to its own instantaneous inertial frame. Show that the rocket's acceleration relation relative to the earth's reference frame at any instant after departure is

$$(1 - v^2/c^2)^{3/2}\, a$$

where v is the earth's velocity relative to the earth at that instant.

Calculate the distance travelled relative to the earth at time t, and, assuming the rocket's acceleration to be maintained indefinitely, show that an appropriately directed light signal emitted from the earth at time t will only overtake the rocket if $t < c/a$.

5

Spin and special relativity

5.1   Angular momentum–a nonrelativistic concept?

During the years following the introduction of quantum mechanics a very successful theory of angular momentum for atomic states was built up, and it played no little part in the explanation of the energy levels of the atom.

The theory is based on the coupling of the orbital ($\mathbf{L}$) and spin ($\mathbf{S}$) angular momentum to yield the total angular momentum $\mathbf{J}$

$$\mathbf{J} = \mathbf{L} + \mathbf{S} \tag{5.1.1}$$

As we shall show later, spin is the intrinsic angular momentum of a body in its own rest frame. Within the context of quantum mechanics $\mathbf{J}$, $\mathbf{L}$ and $\mathbf{S}$ are operators. Their operation on the appropriate quantum mechanical states,*

---

* We shall use the Dirac notation for quantum mechanical systems. For our present purposes this is a convenient mathematical shorthand. The state $|jm\rangle$ can also be regarded as a wave function if the reader is more familiar with that notation.

for example $|jm\rangle$, lead to two eigenvalues

$$\mathbf{J}^2 |jm\rangle = j(j+1) |jm\rangle$$
$$J_z |jm\rangle = m |jm\rangle \qquad (5.1.2)$$

where j and m are the quantum numbers defining the state. Similarly $\mathbf{L}^2 |lm_l\rangle = l(l+1) |lm_l\rangle$ (The Z axis is normally chosen to define the azimuthal quantum number m—the X or Y axis would do equally well).

In what follows we shall not make use of quantum theory. This approach can be justified by Ehrenfest's theorem,[*] which states that the expectation values of quantum mechanical operators behave in the same manner as do the corresponding systems in classical mechanics. Thus it is legitimate to talk about the polarisation (spin) of particles and to describe this property by some (three) vector whose length and direction determine the degree and direction of polarisation respectively, provided that one understands that the system being described represents an averaging over quantum mechanical states. Nevertheless it must be emphasised that the full power and behaviour of the angular momentum states in atomic and nuclear physics can only be properly appreciated within the framework of quantum mechanics.

Equations (5.1.2) raise certain problems. Since we can write

$$\mathbf{J}^2 = J_x^2 + J_y^2 + J_x^2$$

we must ask: what happens to $\mathbf{J}^2$ and J_z when Lorentz transformations are performed? A Lorentz transformation in, say, the Z direction will mix space and time components and the apparent simplicity of equations (5.1.2) might be lost. However, as we stated at the beginning of this section, the theory of angular momentum works perfectly well for atomic systems. If we pause to consider the velocities of the electrons involved in atomic physics we find that they are of order $u \sim 0.01c$; thus, since relativistic effects normally appear as u^2/c^2, atomic systems can be regarded as completely nonrelativistic and the theory of angular momentum, which works so successfully in atomic situations, can be looked upon as the nonrelativistic extension of some more general theory.

5.2 The Pauli–Lubanski vector

We have seen in chapter 4 that the concept of a four-vector in special relativity theory is extremely important, since the scalar product of a four-vector with itself is Lorentz invariant; that is, it remains the same in all reference frames.

We therefore postulate that anything we introduce as a 'relativistic theory of angular momentum' should preferably possess two properties:

(1) it should be Lorentz invariant,

(2) in the limit $u/c \to 0$ it should reduce to the 'normal' theory of angular momentum, since experience has taught us that this theory is completely successful in the domain of atomic physics.

---

[*] EHRENFEST, P., *Z. Phys.* **45**, 455, 1927.

5.2.1 Angular momentum in classical mechanics

Postulate (1) implies that we should firstly construct an appropriate four-vector. Before we do so, let us examine the definition of angular momentum in classical mechanics. It is

$$\mathbf{L} = \mathbf{r} \times \mathbf{p} \qquad (5.2.1)$$

where $\mathbf{L}$ represents the angular momentum generated by a body by its motion (rotation) about a point P which is a distance $|\mathbf{r}|$ from the body and $\mathbf{p}$ is the momentum of the body with respect to the point (figure 5.1). The cross

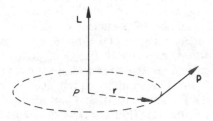

Figure 5.1

in equation (5.2.1) indicates the vector product of $\mathbf{r}$ and $\mathbf{p}$ and so $\mathbf{L}$ is perpendicular to the plane containing $\mathbf{r}$ and $\mathbf{p}$ (compare appendix 2). The components of $\mathbf{L}$ are often written as

$$L_x = yp_z - zp_y$$
$$L_y = zp_x - xp_z \qquad (5.2.2)$$
$$L_z = xp_y - yp_x$$

Now we can use an alternative notation; L_x is concerned with a rotation in the YZ plane and L_y with a rotation in the ZX plane, and so forth. We shall therefore introduce a new notation

$$L_x \equiv L_{yz} \qquad L_y \equiv L_{zx} \qquad L_z \equiv L_{xy} \qquad (5.2.3)$$

This notation has a further advantage; as occasionally we shall have to reverse the order of terms in (5.2.2) we can then write

$$zp_y - yp_z = -L_x = -L_{yz} = L_{zy} \qquad (5.2.4)$$

that is, when we reverse the natural ordering of the sequence xyz, the sign of L changes. L is therefore regarded as an antisymmetric tensor (§4.9), and it is apparent that terms of the type L_{xx} vanish.

The notation of the above section can be summarised in a neat fashion if we revert to our convention of previous chapters of writing

$$x = x_1 \qquad y = x_2 \qquad z = x_3$$

and similarly

$$p_x = p_1 \qquad p_y = p_2 \qquad p_z = p_3$$

Then equations (5.2.1) to (5.2.4) can be summarised in the equation

$$L_j = \sum_{kl} \varepsilon_{jkl} x_k p_l = \tfrac{1}{2} \sum_{kl} \varepsilon_{jkl} L_{kl} \qquad (5.2.5)$$

where ε_{jkl} possesses the following properties

$\varepsilon_{jkl} = +1$ for jkl in the natural sequence for 123; that is

$$\varepsilon_{123} = \varepsilon_{231} = \varepsilon_{312} = +1$$

$\varepsilon_{jkl} = -1$ for jkl not in the natural sequence for 123; that is

$$\varepsilon_{132} = \varepsilon_{213} = \varepsilon_{321} = -1$$

$\varepsilon_{jkl} = 0$ if any two of the subscripts jkl are equal.

As an example of (5.2.5) we may write

$$
\begin{aligned}
L_1 &= \varepsilon_{123} x_2 p_3 + \varepsilon_{132} x_3 p_2 = x_2 p_3 - x_3 p_2 \\
L_1 &= \tfrac{1}{2}(\varepsilon_{123} L_{23} + \varepsilon_{132} L_{32}) \\
&= \tfrac{1}{2}(L_{23} - L_{32}) \\
&= \tfrac{1}{2}(L_{23} + L_{23}) \\
&= L_{23}
\end{aligned}
\qquad (5.2.7)
$$

5.2.2 Spin

The definition

$$\mathbf{L} = \mathbf{r} \times \mathbf{p}$$

of equation (5.2.1) is referred to as the orbital angular momentum of a body about a point P (the distance between the body and P is $|\mathbf{r}|$). In addition, if the body possesses a spin angular momentum $\mathbf{S}$ about some axis in the body, the total angular momentum $\mathbf{J}$ is

$$\mathbf{J} = \mathbf{L} + \mathbf{S} = \mathbf{r} \times \mathbf{p} + \mathbf{S} \qquad (5.2.8)$$

If we assume that our body is point-like, then for $\mathbf{r} = 0$ we are defining $\mathbf{J}$ at the body itself, that is, in the rest frame of the body so that $\mathbf{p} = 0$ and for both $\mathbf{r}$ and $\mathbf{p} = 0$ it is obvious that $\mathbf{L} = 0$. (Moreover, the argument can be extended to finite bodies where the centre of mass is equivalent to the point.) Thus we have

$$\boxed{\mathbf{J} = \mathbf{S} \text{ in the rest frame of the body}} \qquad (5.2.9)$$

and we may define spin as the intrinsic angular momentum of a body in its rest frame.

In the manner of (5.2.2) and (5.2.3) we can introduce spatial components for $\mathbf{J}$

$$J_x = L_x + S_x \text{ and so on}$$

That is

$$J_{yz} = L_{yz} + S_{yz}$$

or in the more general notation of (5.2.5)

$$J_i = \tfrac{1}{2}\sum_{jk}\varepsilon_{ijk}J_{jk} = \tfrac{1}{2}\sum_{jk}\varepsilon_{ijk}(L_{jk}+S_{jk})$$
$$= \sum_{jk}\varepsilon_{ijk}(x_j p_k + \tfrac{1}{2}S_{jk}) \tag{5.2.10}$$

However if we are to deal with relativistic systems we must use four-dimensional notation. We shall therefore write

$$J_{\mu\nu} = L_{\mu\nu} + S_{\mu\nu} \tag{5.2.11}$$

where the subscripts μ, ν run between 1 and 4. In the above relation L is of the form

$$L_{\mu\nu} = x_\mu p_\nu - x_\nu p_\mu$$

and so $L_{\mu\nu}$ is an antisymmetric tensor, that is

$$L_{\mu\nu} = -L_{\nu\mu}$$

(compare (5.2.4) and the remark following). In order to preserve the overall symmetry of equation (5.2.11) we must therefore also postulate that

$$J_{\mu\nu} = -J_{\nu\mu} \qquad S_{\mu\nu} = -S_{\nu\mu} \tag{5.2.12}$$

Our definition of angular momentum in equations (5.2.1) and (5.2.2) implies that spatial components of the angular momentum are generated by rotations in the spatial planes. We must ask what do the rotations in the space-time planes generate. As we have seen in chapters 2 and 4 the answer is associated with the generation of the Lorentz transformations. We shall raise this point again at the end of this chapter.

5.2.3 The Pauli–Lubanski vector*

Let us recall what was said at the beginning of this section—that a 'relativistic theory of angular momentum' should be
(1) Lorentz invariant
(2) reduce to the classical theory in the limit $u/c \to 0$.
In order to satisfy condition (1) a four-vector is required. Now $J_{\mu\nu}$ is based on the product of two four-vectors and so it is a tensor of rank two (compare §4.9). We therefore need to reduce it in some way to a tensor of rank one (a four-vector). This may be done by multiplying it with another four-vector (recall, for example, that the scalar product of a vector with itself yields a

---

* LUBANSKI, J. K., *Physica* 9, 310, 1942.
† The process of pairing up indices is known as saturating or contracting the indices.

tensor of rank zero). Obviously we cannot multiply $J_{\mu\nu}$ by itself to yield a four-vector; so another four-vector is required. The two possible choices are the coordinate, x_ν, and momentum, p_ν, four-vectors[†]

$$A_\mu = \sum_\nu J_{\mu\nu} x_\nu \qquad B_\mu = \sum_\nu J_{\mu\nu} p_\nu$$

The first attempt yields a four-vector which is dependent on space and time—this is not the characteristic of the angular momentum for a body which is not subject to external forces. The second attempt leads to the wrong properties in the limit $u/c \to 0$ where, if we recall (4.2.8)

$$p_\nu \to 0, 0, 0, im_0 c$$

then, for example

$$B_1 \to i J_{14} m_0 c$$

a term which does not appear to have any obvious connection with 'ordinary' angular momentum.

Something more sophisticated must be tried and the solution is the Pauli–Lubanski vector

$$\Gamma_\sigma = \frac{1}{2i} \sum_{\mu\nu\lambda} \varepsilon_{\sigma\mu\nu\lambda} J_{\mu\nu} p_\lambda \qquad (5.2.13)$$

where (compare (5.2.6))

$$\varepsilon_{\sigma\mu\nu\lambda} = +1 \text{ for even permutations of the subscripts 1234}$$
$$= -1 \text{ for odd permutations of 1234} \qquad (5.2.14)$$
$$= 0 \text{ two subscripts equal}$$

Some of the properties of this remarkable vector will be examined in subsequent sections.

5.3   The nonrelativistic limit

We first note that since

$$\Gamma_\sigma = \frac{1}{2i} \sum_{\mu\nu\lambda} \varepsilon_{\sigma\mu\nu\lambda} J_{\mu\nu} p_\lambda$$

and

$$J_{\mu\nu} = L_{\mu\nu} + S_{\mu\nu}$$

then the antisymmetry property of ε implies that

$$\sum_{\mu\nu\lambda} \varepsilon_{\sigma\mu\nu\lambda} L_{\mu\nu} p_\lambda = \sum_{\mu\nu\lambda} \varepsilon_{\sigma\mu\nu\lambda} (x_\mu p_\nu - x_\nu p_\mu) p_\lambda \qquad (5.3.1)$$
$$= 0$$

The proof of this equation follows the manner of (5.2.7) (and of (5.3.3) below), and we leave it as an exercise to the student. This result from (5.3.1) implies that Γ_σ can be rewritten as

$$\Gamma_\sigma = \frac{1}{2i} \sum_{\mu\nu\lambda} \varepsilon_{\sigma\mu\nu\lambda} S_{\mu\nu} p_\lambda \tag{5.3.2}$$

Let us consider the nonrelativistic limit, that is, the rest frame of the body

$$p_\lambda \to 0, 0, 0, im_0 c$$

If we designate Γ by Γ_R in this rest frame we find that

$$\begin{aligned}
\Gamma_{1R} &= \frac{1}{2i} (\varepsilon_{1234} S_{23} + \varepsilon_{1324} S_{32}) \, im_0 c \\
&= \tfrac{1}{2}(S_{23} - S_{32}) \, m_0 c \\
&= S_{23} m_0 c \\
&= S_1 m_0 c
\end{aligned} \tag{5.3.3}$$

since $S_{32} = -S_{23}$ (5.2.12); similarly

$$\Gamma_{2R} = S_2 m_0 c \qquad \Gamma_{3R} = S_3 m_0 c$$

and since $\varepsilon_{4\mu\nu4} = 0$

$$\Gamma_{4R} = 0$$

Thus we see tha the components of Γ_σ are

$$\Gamma_\sigma = S_1 m_0 c, S_2 m_0 c, S_3 m_0 c, 0 \quad \text{for} \quad u/c \to 0 \tag{5.3.4}$$

or in the notation of (4.1.6)

$$\Gamma_R = (\boldsymbol{\Gamma}_R, \Gamma_{4R}) = (m_0 c \mathbf{S}, 0) \tag{5.3.5}$$

and that the Lorentz invariant quantity is

$$\sum_\sigma \Gamma_\sigma^2 = m_0^2 c^2 \mathbf{S}^2 \tag{5.3.6}$$

Therefore, apart from the presence of the term $m_0 c$, it is possible to construct a four-vector which reduces to the normal concept of spin in the limit $u/c \to 0$. The mass term could be avoided by defining a four-vector $\Gamma_R/m_0 c$. However, we shall see in the next section that the form given for Γ in (5.2.13) is necessary in order to obtain sensible results in the limit $u/c \to 1$. Thus we conclude that in the limit $u/c \to 0$ the spatial part of Γ is proportional to $\mathbf{S}$ and that the constant of proportionality is the mass of the particle.

The behaviour of Γ in the nonrelativistic limit (5.3.4) leads to an interesting relation. Since the momentum four-vector for a particle at rest is

$$p = (\mathbf{0}, im_0 c)$$

then

$$\sum_{\sigma=1}^{4} \Gamma_\sigma p_\sigma = 0 \qquad (5.3.7)$$

Now the left-hand side of the above equation is the scalar product of two four-vectors and so the equation is Lorentz invariant. Thus equation (5.3.6) holds not only in the limit $u/c \to 0$, but for all other particle velocities right up to the limit $u/c \to 1$.

5.4   The Lorentz boost

When $u/c \neq 0$, Γ loses some of its simplicity for intermediate velocities, but, as we shall see below, a new simple form emerges in the limit $u/c \to 1$.

In §4.5 it was shown that the Lorentz transformation in an arbitrary direction for a four-vector $A = (\mathbf{A}, A_4)$ was of the form (4.5.6)

$$\mathbf{A}' = \mathbf{A} + \boldsymbol{\beta} \left(\frac{\gamma^2}{\gamma+1} \boldsymbol{\beta} \cdot \mathbf{A} - i\gamma A_4 \right)$$
$$A_4' = \gamma (i\boldsymbol{\beta} \cdot \mathbf{A} + A_4)$$

Now if we consider a particle of mass m_0 at rest and substitute from (5.3.5) $\Gamma_R = (\mathbf{S}m_0 c, 0)$ for A we find that

$$\boldsymbol{\Gamma}' = m_0 c \mathbf{S} + m_0 c \boldsymbol{\beta} \frac{\gamma^2}{\gamma+1} \boldsymbol{\beta} \cdot \mathbf{S}$$
$$\Gamma_4' = i m_0 c \gamma \boldsymbol{\beta} \cdot \mathbf{S}$$

In what follows we shall dispense with the dashes on Γ, since they are not essential once the transformation has been made. The Lorentz boost has also given the particle a three-momentum $\mathbf{p} = m_0 c \gamma \boldsymbol{\beta}$, and so we can rewrite the above equations as

$$\boldsymbol{\Gamma} = m_0 c \mathbf{S} + \frac{\gamma}{\gamma+1} \boldsymbol{\beta} (\mathbf{S} \cdot \mathbf{p})$$
$$= m_0 c \mathbf{S} + \frac{c\mathbf{p}}{E + m_0 c^2} (\mathbf{S} \cdot \mathbf{p}) \qquad (5.4.1)$$
$$\Gamma_4 = i\mathbf{S} \cdot \mathbf{p}$$

where we have inserted $m_0 c^2$ in both top and bottom in the second line and used $E = \gamma m_0 c^2$.

The first of equations (5.4.1) reveals a curious property for $\boldsymbol{\Gamma}$, namely that for 'modest' Lorentz transformations (that is, where $\beta \not\to 1$) the direction of $\boldsymbol{\Gamma}$ does not change.

Consider the plane containing $\boldsymbol{\Gamma}$ and $\boldsymbol{\beta}$ (figure 5.2). If we introduce two unit vectors, $\mathbf{l}$ and $\mathbf{n}$, defined, respectively, parallel and normal to $\boldsymbol{\beta}$

$$\boldsymbol{\beta} = \mathbf{l}\beta \quad \text{where} \quad \beta = |\boldsymbol{\beta}|$$
$$\mathbf{n} \cdot \boldsymbol{\beta} = \mathbf{n} \cdot \mathbf{l}\beta = 0; \quad \mathbf{n}^2 = \mathbf{l}^2 = 1 \qquad (5.4.2)$$

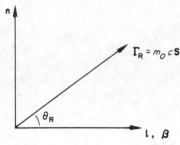

Figure 5.2

and represent the 'length' of S as s

$$\mathbf{S}^2 = s^2$$

then we can write

$$\begin{aligned}\mathbf{\Gamma}_R &= m_0 c\mathbf{S} = m_0 c(\mathbf{l}s \cos\theta_R + \mathbf{n}s \sin\theta_R) \\ &= m_0 cs(\mathbf{l} \cos\theta_R + \mathbf{n} \sin\theta_R)\end{aligned} \tag{5.4.3}$$

where θ_R is the angle between $\mathbf{\Gamma}_R$ and the direction in which the Lorentz transformation $L(\mathbf{\beta})$ is applied (figure 5.2). Now let us apply these considerations to the first of equations (5.4.1) and recall that

$$\mathbf{p} = \gamma m_0 c\mathbf{\beta} = \mathbf{l}\gamma m_0 c\beta$$

we then find that

$$\mathbf{\Gamma} = m_0 cs\left(\mathbf{l} \cos\theta_R + \mathbf{n} \sin\theta_R + \mathbf{l}\frac{\gamma^2\beta^2}{\gamma+1} \cos\theta_R\right)$$

Now

$$\gamma^2\beta^2 = \gamma^2 - 1 = (\gamma-1)(\gamma+1)$$

and so $\mathbf{\Gamma}$ becomes

$$\mathbf{\Gamma} = m_0 cs(\gamma\mathbf{l} \cos\theta_R + \mathbf{n} \sin\theta_R) \tag{5.4.4}$$

Thus we find that $\mathbf{\Gamma}$ differs from $\mathbf{\Gamma}_R$ by the factor γ in front of $\mathbf{l} \cos\theta_R$. Now if $\beta = 0.6$ then $\gamma = 1.25$ and we can see that a Lorentz boost of this strength will not seriously change the direction of $\mathbf{\Gamma}$. For 'modest' Lorentz transformations, that is, for the limit $\beta \to 0$ where Galilean relativity is a good approximation (§3.7), $\gamma \to 1$ and $\mathbf{\Gamma}$ is essentially the same as $\mathbf{\Gamma}_R$. *Thus 'modest' Lorentz transformations leave 'spins' essentially unaltered.* On the other hand if $\beta \to 1$ then $\gamma \to \infty$ and the component parallel to $\mathbf{\beta}$ predominates and $\mathbf{\Gamma}$ points more and more in the direction of $\mathbf{\beta}$, that is, the direction of the momentum of the particle, since

$$\mathbf{p} = \gamma m_0 c\mathbf{\beta} = \mathbf{l}\gamma m_0 u$$

The behaviour of Γ for 'modest' Lorentz transformations can be contrasted with that of the momentum four-vector. For β small it is not difficult to show from (4.5.5) that the momentum vector changes in the manner expected for Galilean transformation

$$\mathbf{p}' - \mathbf{p} \to m_0\beta c \quad \text{for} \quad \beta \to 0$$

so that the spatial direction of the momentum four-vector can change by angles up to $\sim m_0\beta c/|\mathbf{p}|$. If the numerator and denominator are comparable in magnitude the change in direction of $\mathbf{p}$ can then be large.

5.5   The relativistic limit

Now let us consider the relativistic limit $u/c \to 1$ or equivalently $m_0 \to 0$ (compare §§4.3 and 4.4); then $E = c|\mathbf{p}|$ and equations (5.4.1) become

$$\Gamma = \frac{c\mathbf{p}}{c|\mathbf{p}|}(\mathbf{S}\cdot\mathbf{p}) = \frac{\mathbf{S}\cdot\mathbf{p}}{|\mathbf{p}|}\mathbf{p}$$

$$\Gamma_4 = iE\frac{\mathbf{S}\cdot\mathbf{p}}{E} = \frac{\mathbf{S}\cdot\mathbf{p}}{|\mathbf{p}|}p_4 \tag{5.5.1}$$

Thus we find that Γ is proportional to the four-momentum p for a massless particle, and that the constant of proportionality is $\mathbf{S}\cdot\mathbf{p}/|\mathbf{p}|$ which must therefore be Lorentz invariant. Since $\sum_\sigma p_\sigma^2 = 0$ for a particle of mass zero, then

$$\sum_\sigma \Gamma_\sigma p_\sigma = \frac{\mathbf{S}\cdot\mathbf{p}}{|\mathbf{p}|}\sum_\sigma p_\sigma^2 = 0 \tag{5.5.2}$$

and it can be seen that we have recovered the Lorentz invariant equation (5.3.7), which we previously demonstrated for particles at rest.

The term $\mathbf{S}\cdot\mathbf{p}/|\mathbf{p}|$ is called the *helicity* of the particle. It is apparent from (5.5.1) that $\mathbf{S}\cdot\mathbf{p}$ can be a positive or negative quantity depending whether $\mathbf{S}$ is aligned with $\mathbf{p}$ or vice versa. We may therefore also write

$$\Gamma = \pm\lambda\mathbf{p} \quad \text{where} \quad \lambda = \frac{|\mathbf{S}\cdot\mathbf{p}|}{|\mathbf{p}|}$$

Particles in $+$ or $-$ states are said to have positive or negative helicity respectively.

Thus particles with zero rest mass have only two directions of polarisation, no matter how large their spin is. (The intimate connection of the neutrino (a particle of mass zero) with the weak interactions of elementary particle physics leads to the result that only one of the helicity states $\pm\lambda$ is occupied.*) This behaviour contrasts with the $2S+1$ directions of polarisation for particles of nonzero rest mass and spin S found in nonrelativistic quantum mechanics.

The alignment of spin with momentum can, of course, occur for any

---

* Only two particles with zero mass are at present known—the photon (two helicity states) and neutrino (one helicity state).

particle of finite mass—*provided we choose the appropriate reference frame.* However, once a Lorentz transformation is made in a direction not parallel to **p**, the alignment is lost. Only when the particle has zero mass so that the conditions (5.5.1) are fulfilled will the spin (that is, **Γ**) follow the momentum **p** in every reference frame.

If we connect internal motion of the body with its spin, then this motion is perpendicular to the direction of **p** for massless particles (compare figure 5.1 with **S** substituted for **L**). Right-handed motion ('right-handed circular polarisation') is normally associated with spin pointing parallel to **p** in elementary particle physics and vice versa for left-handed polarisation. We shall return to the topic of helicity in chapter 6.

5.6   Symmetries in physics

5.6.1   Symmetries and the interval

Earlier in this chapter we have mentioned spatial rotations and associated them with angular momentum. In chapters 2 to 4 we have repeatedly connected rotations in space-time with Lorentz transformations. In this section these statements will be extended a little further.

Let us recall our definition of the interval between two events a and b (2.4.1)

$$S_{ab}^2 = c^2(t_a - t_b)^2 - (x_a - x_b)^2 - (y_a - y_b)^2 - (z_a - z_b)^2 \qquad (5.6.1)$$

Now we can make certain changes to the right-hand side of this equation and leave S_{ab}^2 unaltered, that is, invariant. They are:

(1) displacements in space—if for example we displace the XYZ coordinate frame along the X axis by an amount $-\delta$ so that x_a and x_b become

$$x_a' = x_a + \delta \qquad x_b' = x_b + \delta \qquad (5.6.2)$$

then

$$(x_a' - x_b')^2 = (x_a - x_b)^2 \qquad (5.6.3)$$

and the argument can be extended to all three spatial axes

(2) displacements in time—the same considerations apply as for (5.6.2) and (5.6.3)

(3) rotations in space-time—these were discussed in §2.6

(4) rotations in space—the basic geometry is the same as in (3)

(5) reflections of spatial coordinates—if we simultaneously reflect X, Y and Z axes then x_a becomes $x_a' = -x_a$ and so forth, but $(x_a' - x_b')^2 = (x_a - x_b)^2$

(6) reflections of time coordinates—the same considerations apply as for (5).

The above operations are all examples of *symmetries*—briefly a system is symmetrical if after some operation to it, it appears to be the same as before. Examples of systems possessing symmetry are a ball bearing whose appearance remains the same after a rotation, and a sugar cube which appears the same after rotation through 90°. These are examples of geometrical symmetries—a subject of considerable importance in crystallography.

Probably the most extensive exploitation of symmetry principles occurs in high-energy particle physics, and it was the original study of the relativity problem that led to a concentration of effort on symmetry in the laws of physics.

A proper appreciation of symmetry principles at atomic distances can only be formulated with the aid of quantum mechanics—and that subject is outside the scope of this book. Nevertheless we can get some inkling of what is involved from our present studies. We have already seen that the notion of the invariance of the interval under rotations in space-time lead to the equations of the Lorentz transformation and the principle that the scalar product of four-vectors is a Lorentz invariant—that is, it remains invariant under Lorentz transformations. This is an example of a *conservation law*—namely that the scalar product of four-vectors remains conserved under Lorentz transformations, that is the symmetry associated with rotations in space-time. Let us sketch another. For constant spatial displacements velocity remains unaltered (compare (5.6.2))

$$u' = \frac{dx'}{dt} = \frac{d}{dt}(x+\delta) = \frac{dx}{dt} = u \tag{5.6.4}$$

and since momentum $= mu$, then, if conservation of momentum in collisions of bodies holds in a reference frame XYZ, it will also hold in the spatially displaced frame $X'Y'Z'$. However, if the theory is properly formulated* one can go even further and show that invariance of a system under spatial displacement leads to conservation of momentum in any reference frame.

The complete examination of symmetry within the framework of quantum mechanics leads to a remarkable law—that *for each symmetry principle there is a corresponding conservation law*. In the accompanying table we list the symmetries associated with the invariance of the interval (5.6.1).

Table 5.1

Symmetry	Conservation Law
Spatial displacement	Linear momentum
Temporal displacement	Energy
Space-time rotation	Lorentz invariance
Spatial rotation	Angular momentum
Spatial reflection	Parity
Temporal reflection	Time reversal

The list given in the table corresponds to the symmetries associated with invariance of the interval only—many other symmetries exist in quantum mechanics.

5.6.2   The violation of the conservation laws

The most interesting symmetries listed in table 5.1 are those concerning

* The interested reader can find elementary discussions at the quantum mechanical level in the present author's *Notes on Elementary Particle Physics* pages 19 and 37 and at the field theoretical level in his *Elementary Particle Physics* page 88 and 183, Pergamon Press.

spatial and temporal reflection, since in recent years it has been discovered that parity conservation is definitely violated in many processes in elementary particle physics, and time reversal invariance is in doubt in one particular process.

The processes occurring at the atomic and subatomic level of elementary particle physics divide into three classes, depending on the strength of interaction of the particles with each other. In order of decreasing strength these classes are the strong, electromagnetic and weak interactions. In quantum mechanical language, physical processes are described in terms of amplitudes, A, and the square of the moduli of these amplitudes $|A|^2$ represents the probability for the process to occur. If we use a scale of 1 for strong interactions, then experiments have given the following relative strength for the three categories of interactions

$$
\begin{aligned}
&\text{strong interactions} &&|A|^2 \sim 1 \\
&\text{electromagnetic interactions} &&|A|^2 \sim 10^{-2} \\
&\text{weak interactions} &&|A|^2 \sim 10^{-10}
\end{aligned}
\tag{5.6.5}
$$

The strong interactions are responsible for the tight binding of the protons and neutrons inside nuclei, whilst atomic interactions (which always involve photons either directly or indirectly) are examples of electromagnetic interactions; nuclear β-decay provides an example of a weak interaction. The word 'weak' is only used relatively; the 'weak interactions' of elementary particles are far stronger than their gravitational interactions.

All the weak interactions are found to violate the conservation of parity. The word 'parity' can only be properly formulated within the framework of quantum mechanics (indeed it has no meaning in classical mechanics). For our present purposes we shall not attempt a definition but merely note that if parts of the amplitude A change sign upon reflection of spatial coordinates then conservation of parity is violated.

Consider the behaviour of simple terms under spatial reflection

$$
\begin{aligned}
&\text{Spatial coordinates } \mathbf{r} \xrightarrow{\text{Reflection}} &&\mathbf{r}' = -\mathbf{r} \\
&\text{Linear momentum } \mathbf{p} = m\frac{d\mathbf{r}}{dt} &&\mathbf{p}' = -\mathbf{p} \\
&\text{Orbital angular momentum } \mathbf{L} = \mathbf{r} \times \mathbf{p} &&\mathbf{L}' = -\mathbf{r} \times -\mathbf{p} \\
& &&= \mathbf{L} \\
&\text{Spin } \mathbf{S} &&\mathbf{S}' = \mathbf{S}
\end{aligned}
\tag{5.6.6}
$$

where the behaviour of spin follows that for $\mathbf{L}$. (We can always consider the spin arising from the internal rotation of the body).

Let us attempt to construct an amplitude for a nuclear process (say, β-decay); if we postulate it to consist of two separate amplitudes a and b which are independent and dependent on $\mathbf{S} \cdot \mathbf{p}$ respectively

$$
A = a + b\mathbf{S} \cdot \mathbf{p}
\tag{5.6.7}
$$

then the transition probability is

$$|A|^2 = |a + b\mathbf{S} \cdot \mathbf{p}|^2$$

Now if we consider a spatial reflection, we find

$$A \to A' = a - b\mathbf{S} \cdot \mathbf{p}$$

by virtue of (5.6.6). Thus if the amplitude conserves parity, so that $A = A'$, then b must equal zero. If $b \neq 0$ then effects indicating an alignment of spin and momentum could exist.

An analysis by Lee and Yang[*] in 1956 of the behaviour of strong, electromagnetic and weak interactions led to the conclusion that whilst there was good evidence to support the assumption of conservation of parity in strong and electromagnetic interactions no conclusions could be drawn about weak interactions. They suggested tests for the parity conservation hypothesis by examining the correlation between spin (that is, polarisation) and momentum of outgoing particles in weak interactions. The suggestions of Lee and Yang were put to experimental test in nuclear β-decay[†] and muon decay[‡] and strong spin–momentum correlation was found (in both cases the results indicated $a \sim b$). Later work revealed an interesting relativistic effect, the detailed theory of the β-decay process predicted that the strength of the polarisation along the direction of the momentum of the outgoing electron in β-decay should be proportional to u/c where u represents the velocity of the electron in the laboratory reference frame. This effect was confirmed.

At the beginning of this section we mentioned that time reversal invariance is in doubt for one particular process. The story is complicated and we shall not pursue it here—it apparently applies to one process only (the decay of K^0 mesons) and in contrast to the strong violation of parity in weak interactions ($a \sim b$ in (5.6.7)), the time violation is weak (that is, $b \ll a$ in the language of (5.6.7)). The weakness of the violation has led to the present uncertainty concerning whether the effect occurs at all.

PROBLEMS

5.1 Prove equation (5.3.1).

5.2 In the rest frame (Σ) of a particle of mass m its 'spin', $\mathbf{\Gamma_R}$, points at 45° to the X axis in the XY plane. Where does it point after successive transformations corresponding to $\beta = 0.8$ and 0.6 along the X and Y axes respectively? Where does the momentum vector $\mathbf{p}$ also point after these transformations. By how much does the direction of $\mathbf{\Gamma}$ and $\mathbf{p}$ change between the first and second transformations?

---

[*] Lee, T. D., and Yang, C. N., *Phys. Rev.* **104**, 254, 1956.
[†] Wu, C. S., Ambler, E., Hayward, R. W., Hoppes, D. D., and Hudson, R. P., *Phys. Rev.* **105**, 1413, 1957.
[‡] Friedman, J. I., and Telegdi, V. L., *Phys. Rev.* **105**, 1681, 1957. Garwin, R. L., Lederman, L. M. and Weinrich, M., *Phys. Rev.* **105**, 1415, 1957.

5.3* In the decay process

$$\pi^- \to \mu^- \bar{\nu}$$

the muons (μ^-) are born with their spins longitudinally polarised; that is, their spins point along their direction of motion. We illustrate this point by the following diagram, drawn in the nonrelativistic limit; in it $v_\mu(\pi)$ represents the velocity of the muon in the π rest frame.

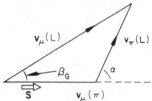

If the pion is itself moving in the laboratory frame with a velocity $v_\pi(L)$, the final velocity of the muon is $v_\mu(L)$ in the laboratory frame. It is evident that if we go to the muon rest frame by slowing down the muon along the direction $v_\mu(L)$ the muon polarisation direction will be signified by β_G. Show that in the nonrelativistic (Galilean) limit

$$\tan\beta_G = \frac{\tan\alpha}{1 + v_\mu(\pi)/v_\pi(L)\cos\alpha}$$

5.4 In the relativistic situation the triangle of the previous equation becomes a spherical triangle.

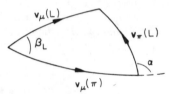

Using the geometry of spherical triangles (appendix 3) or the equations of the Lorentz transformation show that

$$\tan\beta_L = \sqrt{[1 - v_\mu^2(\pi)/c^2]}\tan\beta_G$$

5.5 Where does the muon polarisation point in the laboratory frame in the previous question if $v_\pi(L)$ is chosen as the reference axis?

5.6 In chapter 4 the relation $\sum_\lambda p_\lambda^2 = -m_0^2 c^2$ for the momentum four vector was used to show that the positions of the momentum four vector could be represented on a sphere of radius $im_0 c$. Show that the relation $\sum_\lambda \Gamma_\lambda^2 = m_0^2 c^2 S^2$ for the Pauli–Lubanski four-vector implies that it also can be represented by a sphere, but whereas the physical limits for θ for p are 0 to 45°, in the notation of figure 4.5 they are 45° to 90° for Γ_λ. Show that for a particle with velocity $u \to c$ the angles describing the position of the radius vector coincide for p_λ and Γ_λ.

5.7 Show that an element of volume in 'momentum space'

$$\frac{dp_x\, dp_y\, dp_z}{E}$$

* Questions 5.3 and 5.4 are based on the CERN internal report SC/13113/Rapp./142 (1960) by V. L. Telegdi.

is invariant under Lorentz transformations in the X direction. Show that the above term is invariant under spatial rotations, and hence it is invariant under Lorentz transformations in arbitrary directions.

5.8 Calculate the ratio of the forces of gravitational attraction to the coulomb repulsion for two protons separated by a distance r. (Charge on proton $= 1.6 \times 10^{-19}$ coulomb; mass of proton $= 1.6 \times 10^{-27}$ kg; gravitational constant $= 6.7 \times 10^{-11}$ m^3 kg^{-1} s^{-2}).

6

The Lorentz invariance of physical theories

In the previous chapters considerable stress has been laid on the description of physical processes in a Lorentz invariant manner. This result was achieved by the use of the four-vector, whose scalar product with itself or other four-vectors formed a Lorentz invariant quantity. Up to now the four-vectors we have examined are those which describe the mechanical properties of physical systems—coordinates, momenta and so forth. In this chapter we shall extend this treatment to some of the basic dynamical equations which describe physical systems, with especial emphasis on Maxwell's equations.

6.1 Maxwell's equations and the conservation of charge

6.1.1 The equation of continuity

Maxwell's equations, *in* MKS *units*, possess the following form for an assembly of charges in free space*

$$\mathbf{V} \times \mathbf{B} = \mu_0 \left(j + \varepsilon_0 \frac{\partial \mathbf{E}}{\partial t} \right)$$

$$\mathbf{V} \cdot \mathbf{B} = 0$$

$$\mathbf{V} \times \mathbf{E} = -\frac{\partial \mathbf{B}}{\partial t} \qquad (6.1.1)$$

$$\mathbf{V} \cdot \mathbf{E} = \frac{\varrho}{\varepsilon_0}$$

where

$$\mathbf{V} \times = \text{curl (appendix 2)}$$
$$\mathbf{V} \cdot = \text{div (appendix 2)}$$
$$\mathbf{B} = \text{magnetic field}$$
$$\mathbf{E} = \text{electric field}$$
$$\mathbf{j} = \text{current density}$$
$$\varrho = \text{charge density}$$
$$\varepsilon_0 = \text{electric space constant}; 8.8 \times 10^{-12} \text{ farad m}^{-1}$$
$$\mu_0 = \text{magnetic space constant}; 4 \times 10^{-7} \text{ henry m}^{-1}$$

and

$$\mu_0 \varepsilon_0 = 1/c^2$$
$$c = \text{velocity of light} \qquad (6.1.2)$$

Our objective is to get these equations into a form which has more connection with the four-vectors we have been discussing so far. Since div·curl equals zero (appendix 2), we may rewrite the first of equations (6.1.1) and combine it with the last to obtain

$$\mathbf{V} \cdot \mathbf{V} \times \mathbf{B} = 0$$

$$= \mu_0 \left(\mathbf{V} \cdot \mathbf{j} + \varepsilon_0 \frac{\partial}{\partial t} \mathbf{V} \cdot \mathbf{E} \right)$$

$$= \mu_0 \left(\mathbf{V} \cdot \mathbf{j} + \frac{\partial \varrho}{\partial t} \right)$$

The result

$$\mathbf{V} \cdot \mathbf{j} + \frac{\partial \varrho}{\partial t} = 0 \qquad (6.1.3)$$

or

$$\frac{\partial j_x}{\partial x} + \frac{\partial j_y}{\partial y} + \frac{\partial j_z}{\partial z} + \frac{\partial \varrho}{\partial t} = 0 \qquad (6.1.4)$$

---

* No attempt will be made to discuss electromagnetism in material media in this book.

is known as the *equation of continuity*. Its physical significance will be discussed later; at present we shall concentrate on producing a four-vector notation.

If equation (6.1.4) is rewritten as

$$\frac{\partial j_x}{\partial x}+\frac{\partial j_y}{\partial y}+\frac{\partial j_z}{\partial z}+\frac{ic}{ic}\frac{\partial \varrho}{\partial t}=0$$

the denominators represent the elements of a four-vector. Let us therefore write

$$x=x_1 \quad y=x_2 \quad z=x_3 \quad ict=x_4$$

and introduce a four-current with components

$$j_x=j_1 \quad j_y=j_2 \quad j_z=j_3 \quad ic\varrho=j_4 \tag{6.1.5}$$

then equation (6.1.3) becomes

$$\mathbf{V}\cdot\mathbf{j}+\frac{\partial \varrho}{\partial t}=\sum_{\lambda=1}^{4}\frac{\partial j_\lambda}{\partial x_\lambda}=0 \tag{6.1.6}$$

Thus the elements of $\mathbf{j}$ and ϱ appear to be associated with a four-vector j_λ. The equation (6.1.6) is in fact sufficient to establish that j behaves like a four-vector under Lorentz transformations. However, rather than examine the formal manipulations, let us make some simple physical arguments. Current is defined as the total charge crossing a given element of area per second. Thus the current density is given by the product of the charge density and its velocity of flow. Now for a system of charges at rest we can designate the charge density by ϱ_0 and the current density will be zero since the velocity is zero, hence if j_λ is a four-vector then by (6.1.5)

$$\sum_{\lambda=1}^{4}j_\lambda^2=-c^2\varrho_0^2 \tag{6.1.7}$$

Now we have previously encountered a four-vector the sum of whose components squared added up to $-c^2$. It was the four-velocity U of § 4.2 (compare (4.2.3) and (4.2.5))

$$U_\lambda=c\frac{dx_\lambda}{dS}=\gamma_u\frac{dx_\lambda}{dt}$$

$$\sum_{\lambda=1}^{4}U_\lambda^2=-c^2$$

Thus we can make the identification

$$j_\lambda=\varrho_0 U_\lambda \tag{6.1.8}$$

and furthermore

$$\sum_{\lambda=1}^{4}\frac{\partial j_\lambda}{\partial x_\lambda}=\varrho_0\sum_{\lambda=1}^{4}\frac{\partial U_\lambda}{\partial x_\lambda}=\varrho_0 c\frac{\partial}{\partial S}\sum_{\lambda=1}^{4}\frac{\partial x_\lambda}{\partial x_\lambda}=0$$

and so the equation of continuity (6.1.6) is satisfied.

We have thus established that j_λ is related to U_λ by a constant of proportionality ϱ_0 (the charge density in the stationary system), and since U_λ behaves like

$$U'_\lambda = \sum_\alpha L_{\lambda\alpha} U_\alpha$$

under Lorentz transformations (compare (4.2.4)), then

$$j'_\lambda = \sum_\alpha L_{\lambda\alpha} j_\alpha \qquad (6.1.9)$$

6.1.2   Charge conservation in Lorentz transformations

The total charge Q in a system of volume V is given by

$$Q = \int \varrho \, dV \qquad (6.1.10)$$

where the charge density ϱ and V are determined by the same observer.

If the charge system is at rest relative to the observer, then the charge contained in an element of volume dV_0 is

$$dQ_0 = \varrho_0 dV_0 = \varrho_0 \, dx \, dy \, dz$$

where dx, dy and dz are defined in the observer's rest frame. For motion of the system in the X direction at a velocity u the volume element becomes

$$dV = dV_0 \sqrt{(1 - u^2/c^2)}$$

to an external observer by virtue of the equation of the Lorentz contraction (3.1.7). At the same time the charge density will change. At rest, the elements j are given by

$$j_\lambda = 0, 0, 0, i\varrho_0 c$$

as we have seen in the previous section. The Lorentz transformation relation (6.1.9)

$$j'_\lambda = \sum_\alpha L_{\lambda\alpha} j_\alpha$$

then reduces for the fourth component, (with the aid of (4.1.9)) to

$$j'_4 = ic\varrho = L_{44} \, ic\varrho_0 = ic\gamma\varrho_0 = \frac{ic\varrho_0}{\sqrt{(1 - u^2/c^2)}}$$

and so we have

$$\varrho = \frac{\varrho_0}{\sqrt{(1 - u^2/c^2)}} = \gamma_u \varrho_0 \qquad (6.1.11)$$

and therefore the charge element dQ_0 remains invariant

$$dQ_0 = \varrho_0 dV_0 = \varrho \, dV = dQ \tag{6.1.12}$$

Upon summing (or integrating) over the elements dV it is apparent that the total charge Q (6.1.10) will also remain constant under a Lorentz transformation.

This result might be expected from physical intuition. If we had N charges, each of magnitude e, confined in a box of volume V_0 so that $Q = Ne$, we would expect the distance of separation of the charges to change under a Lorentz transformation (that is ϱ to change), but the transformation produces no obvious physical mechanism to cause the changes to flow out of the box, and so the count of N charges should remain unaltered.

Let us finally return to the physical interpretation of the equation of continuity itself (6.1.3). If this equation is integrated over the volume of any container and Gauss' theorem is used, then we obtain

$$\frac{\partial}{\partial t} \int \varrho \, dV = \frac{\partial Q}{\partial t} = -\int \nabla \cdot \mathbf{j} \, dV = -\int j_n \, dA \tag{6.1.13}$$

This equation represents the statement that the rate of loss of charge from a system of volume V equals the total flow of current out from the surface area A (here j_n represents the current normal to the surface element dA). The physical significance of the minus sign in (6.1.13) is left as an exercise for the reader.

In conclusion it is worth noting one final property of the current and charge densities. The relation (6.1.8)

$$j_\lambda = \varrho_0 U_\lambda$$

in conjunction with the definition of U_λ (4.2.3), implies that

$$j_\lambda = \gamma_u \varrho_0 \frac{dx_\lambda}{dt}$$

We have already seen from (6.1.11) that $\varrho = \gamma_u \varrho_0$, hence the components of j_λ are

$$j_\lambda = \varrho \, \frac{dx_1}{dt}, \varrho \, \frac{dx_2}{dt}, \varrho \, \frac{dx_3}{dt}, ic\varrho$$
$$= \varrho u_1, \quad \varrho u_2, \quad \varrho u_3, \quad ic\varrho$$

or in the notation of (4.1.6)

$$j = (\varrho \mathbf{u}, ic\varrho) \tag{6.1.14}$$

That is the current density is given by the product of the charge density and velocity in every reference system.

6.2 Maxwell's equations and the Lorentz transformation

6.2.1 Reformulation of the equations

The Maxwell equations given in (6.1.1) have no obvious connection with Lorentz invariance or even four-vectors. Our first task must therefore be to construct a more suitable form. This can be done by introducing the vector, **A**, and scalar, ϕ, potentials of electromagnetic theory. These potentials have no physical significance, but they possess considerable mathematical importance as auxiliary functions which make the task of handling the equations considerably easier.

The first and last of Maxwell's equations can be rewritten as

$$\mathbf{V} \times \mathbf{B} - \frac{1}{c^2}\frac{\partial \mathbf{E}}{\partial t} = \mu_0 \mathbf{j}$$

$$ic\mu_0\varepsilon_0\, \mathbf{V} \cdot \mathbf{E} = ic\mu_0\varrho = \mu_0 j_4 \tag{6.2.1}$$

and since the right-hand sides transform like the space and time components of a common four-vector, then the left-hand sides must also possess the same property.

Consider the second of Maxwell's equations

$$\mathbf{V} \cdot \mathbf{B} = 0$$

Since div·curl is zero (appendix 2) this equation implies that **B** is the curl of some other term. We therefore write

$$\mathbf{B} = \mathbf{V} \times \mathbf{A} \tag{6.2.2}$$

and hence

$$\mathbf{V} \cdot \mathbf{B} = \mathbf{V} \cdot \mathbf{V} \times \mathbf{A} = 0$$

Next let us introduce (6.2.2) into the third of Maxwell's equations, then

$$\mathbf{V} \times \mathbf{E} = -\frac{\partial \mathbf{B}}{\partial t}$$

becomes with the aid of (6.2.2)

$$\mathbf{V} \times \left(\mathbf{E} + \frac{\partial \mathbf{A}}{\partial t}\right) = 0$$

This equation is satisfied if the system in brackets is the gradient of a scalar, since curl × grad is zero. We therefore write

$$\mathbf{E} + \frac{\partial \mathbf{A}}{\partial t} = -\mathbf{V}\phi$$

or

$$E = -\nabla\phi - \frac{\partial A}{\partial t} \tag{6.2.3}$$

Now there is a curious ambiguity about these definitions. If we make the simultaneous changes

$$A \to A' = A + \nabla\chi$$
$$\phi \to \phi' = \phi - \frac{\partial\chi}{\partial t} \tag{6.2.4}$$

where χ is some arbitrary function, then **B** and **E** are left unchanged—and recall that it is the magnetic and electric fields which are accessible to experiment in electromagnetic systems, not **A** and ϕ. The proof of the invariance of **B** and **E** under the changes (6.2.4) is readily made

$$B' = \nabla \times A' = \nabla \times A + \nabla \times \nabla\chi$$
$$= \nabla \times A$$
$$= B$$

$$E' = -\nabla\left(\phi - \frac{\partial\chi}{\partial t}\right) - \frac{\partial}{\partial t}(A + \nabla\chi)$$
$$= -\nabla\phi - \frac{\partial A}{\partial t}$$
$$= E$$

The invariance of **B** and **E** for arbitrary values of **A** and ϕ is known as *gauge invariance*. The arbitrary nature of **A** and ϕ can be exploited to introduce the *Lorentz gauge* or *condition*, the purpose of which will become apparent below. This condition is

$$\nabla \cdot A + \frac{1}{c^2}\frac{\partial\phi}{\partial t} = 0 \tag{6.2.5}$$

Now let us introduce a new definition

$$A_4 = \frac{i}{c}\phi \tag{6.2.6}$$

then equation (6.2.5) becomes (compare (6.1.6))

$$\nabla \cdot A + \frac{i}{ic^2}\frac{\partial\phi}{\partial t} = \nabla \cdot A + \frac{\partial A_4}{\partial x_4} = \sum_{\lambda=1}^{4}\frac{\partial A_\lambda}{\partial x_\lambda} = 0 \tag{6.2.7}$$

Thus we see that **A** and $i\phi/c$ appear to behave like the space and time components of a four-vector. This conclusion can be reinforced when we examine the remaining Maxwell equations. If we again make use of appendix 2

$$\mathbf{V} \times \mathbf{V} \times \mathbf{A} = \mathbf{V}\mathbf{V} \cdot \mathbf{A} - \nabla^2 \mathbf{A}$$

the first and fourth of Maxwell's equations (6.2.1) become (with the aid of (6.2.7))

$$
\begin{aligned}
\mathbf{V} \times \mathbf{B} - \frac{1}{c^2}\frac{\partial \mathbf{E}}{\partial t} &= \mathbf{V} \times \mathbf{V} \times \mathbf{A} - \frac{1}{c^2}\frac{\partial}{\partial t}\left(-\mathbf{V}\phi - \frac{\partial \mathbf{A}}{\partial t}\right) \\
&= -\nabla^2 \mathbf{A} + \frac{1}{c^2}\frac{\partial^2 \mathbf{A}}{\partial t^2} + \mathbf{V}\left(\mathbf{V}\cdot\mathbf{A} + \frac{1}{c^2}\frac{\partial \phi}{\partial t}\right) \\
&= -\nabla^2 \mathbf{A} + \frac{1}{c^2}\frac{\partial^2 \mathbf{A}}{\partial t^2} \\
&= \mu_0 \mathbf{j}
\end{aligned}
\tag{6.2.8}
$$

$$
\begin{aligned}
ic\mu_0\varepsilon_0 \mathbf{V}\cdot\mathbf{E} &= \frac{i}{c}\mathbf{V}\cdot\left(-\mathbf{V}\phi - \frac{\partial \mathbf{A}}{\partial t}\right) \\
&= -\frac{i}{c}\nabla^2\phi - \frac{i}{c}\frac{\partial}{\partial t}\mathbf{V}\cdot\mathbf{A} \\
&= -\nabla^2 A_4 + \frac{1}{c^2}\frac{\partial^2 A_4}{\partial t^2} \\
&= \mu_0 j_4
\end{aligned}
\tag{6.2.9}
$$

Thus we may summarise these equations as

$$
\begin{aligned}
\nabla^2 \mathbf{A} - \frac{1}{c^2}\frac{\partial^2 \mathbf{A}}{\partial t^2} &= -\mu_0 \mathbf{j} \\
\nabla^2 A_4 - \frac{1}{c^2}\frac{\partial^2 A_4}{\partial t^2} &= -\mu_0 j_4
\end{aligned}
\tag{6.2.10}
$$

which can be combined to yield

$$\nabla^2 A_\alpha - \frac{1}{c^2}\frac{\partial^2 A_\alpha}{\partial t^2} = \sum_{\lambda=1}^{4}\frac{\partial^2 A_\alpha}{\partial x_\lambda^2} = -\mu_0 j_\alpha \tag{6.2.11}$$

Thus the use of the Lorentz gauge condition (6.2.5) has reduced Maxwell's equations to a very compact and simple form, and in subsequent sections we shall examine some of the implications of these equations.

6.2.2   The Lorentz transformation of Maxwell's equations

In putting Maxwell's equations into a suitable notation, we have obtained three important equations:

(1) the equation of continuity (6.1.6)

$$\sum_\lambda \frac{\partial j_\lambda}{\partial x_\lambda} = 0 \tag{6.2.12}$$

(2) the Lorentz condition (6.2.7)

$$\sum_\lambda \frac{\partial A_\lambda}{\partial x_\lambda} = 0 \tag{6.2.13}'$$

(3) Maxwell's equations (6.2.11)

$$\sum_\lambda \frac{\partial^2 A_\alpha}{\partial x_\lambda^2} = -\mu_0 j_\alpha \tag{6.2.14}$$

Our task is to verify that these equations are Lorentz invariant, that is, that they remain unaltered in form under Lorentz transformations, for example

$$\sum_{\lambda=1}^{4} \frac{\partial j'_\lambda}{\partial x'_\lambda} = \sum_{\lambda=1}^{4} \frac{\partial j_\lambda}{\partial x_\lambda}$$

All three equations possess ∂x in the denominators, and so let us recall the rules for partial differentiation in algebra; we can write

$$\frac{\partial}{\partial x'_1} = \frac{\partial x_1}{\partial x'_1} \frac{\partial}{\partial x_1} + \frac{\partial x_2}{\partial x'_1} \frac{\partial}{\partial x_2} + \frac{\partial x_3}{\partial x'_1} \frac{\partial}{\partial x_3} + \frac{\partial x_4}{\partial x'_1} \frac{\partial}{\partial x_4} \tag{6.2.15}$$

and for a Lorentz transformation in the X direction the standard transformation equations (2.6.15) can be used to obtain

$$\frac{\partial}{\partial x'_1} = \gamma \frac{\partial}{\partial x_1} + i\beta\gamma \frac{\partial}{\partial x_4} = \gamma \frac{\partial}{\partial x_1} + i\frac{v}{c}\gamma \frac{\partial}{\partial x_4} \tag{6.2.16}$$

The same procedure could be adopted to obtain $\partial/\partial x'_2$ and so on; however, it will be convenient to introduce the shorthand based on matrices (compare §4.1) for future use.

The standard Lorentz transformation in the X direction given in (2.6.14) can be represented, in the notation of (4.1.10), as

$$x'_\lambda = \sum_{\sigma=1}^{4} L_{\lambda\sigma} x_\sigma \tag{6.2.17}$$

where

$$\lambda \rightarrow$$
$$L = \begin{bmatrix} \gamma & 0 & 0 & i\beta\gamma \\ 0 & 1 & 0 & 0 \\ 0 & 0 & 1 & 0 \\ -i\beta\gamma & 0 & 0 & \gamma \end{bmatrix} \begin{matrix} \sigma \\ \downarrow \end{matrix} \tag{6.2.18}$$

similarly the reverse transformation (2.6.15) becomes (compare (4.1.12))

$$x_\lambda = \sum_{\sigma=1}^{4} L_{\sigma\lambda} x'_\sigma \tag{6.2.19}$$

Whilst this notation is perfectly acceptable, the rules of matrix algebra permit us to write the elements of the inverse of a nonsingular square matrix with determinant one as

$$L_{\sigma\lambda} = L_{\lambda\sigma}^{-1} \qquad (6.2.20)$$

and it is in this form that the inverse transformation is normally given. Thus equation (6.2.19) can also be represented as

$$x_\lambda = \sum_{\sigma=1}^{4} L_{\lambda\sigma}^{-1} x_\sigma \qquad (6.2.21)$$

where

$$L^{-1} = \begin{bmatrix} \gamma & 0 & 0 & -i\beta\gamma \\ 0 & 1 & 0 & 0 \\ 0 & 0 & 1 & 0 \\ i\beta\gamma & 0 & 0 & \gamma \end{bmatrix} \begin{matrix} \\ \sigma \\ \\ \downarrow \end{matrix} \qquad (6.2.22)$$

It is not difficult to show that the multiplication of the elements of the two matrices L and L^{-1} lead to a very simple result

$$\sum_{\sigma=1}^{4} L_{\lambda\sigma} L_{\sigma\alpha}^{-1} = \sum_{\sigma=1}^{4} L_{\lambda\sigma}^{-1} L_{\sigma\alpha} = \delta_{\lambda\alpha} \qquad (6.2.23)$$

where δ represents the Kroenecker delta symbol

$$\begin{aligned} \delta &= 1 \quad \text{for} \quad \lambda = \sigma \\ \delta &= 0 \quad \text{for} \quad \lambda \neq \sigma \end{aligned} \qquad (6.2.24)$$

As an example of the use of the above equations, let us recall equation (4.1.5)

$$\sum_{\lambda=1}^{4} A_\lambda' B_\lambda' = \sum_{\lambda=1}^{4} A_\lambda B_\lambda$$

then (all indices run from 1 to 4)

$$\begin{aligned} \sum_\lambda A_\lambda' B_\lambda' &= \sum_{\lambda\alpha\beta} L_{\lambda\alpha} L_{\lambda\beta} A_\alpha B_\beta \\ &= \sum_{\lambda\alpha\beta} L_{\alpha\lambda}^{-1} L_{\lambda\beta} A_\alpha B_\beta \\ &= \sum_{\alpha\beta} \delta_{\alpha\beta} A_\alpha B_\beta \\ &= \sum_\alpha A_\alpha B_\alpha \end{aligned}$$

and since both λ and α run over the same indices (1 to 4) equation (4.1.5) is proved. The utility of using the matrix elements as a shorthand can be appreciated when one realises that the first line on the right-hand side of the above equations contains 64 terms.

Let us now return to equation (6.2.15); it can be rewritten as

$$\frac{\partial}{\partial x'_1} = \sum_{\lambda=1}^{4} \frac{\partial x_\lambda}{\partial x'_1} \frac{\partial}{\partial x_x} = \sum_{\lambda=1}^{4} L_{x1}^{-1} \frac{\partial}{\partial x_\lambda}$$

and it is evident that in general

$$\frac{\partial}{\partial x'_\sigma} = \sum_{\lambda=1}^{4} L_{\lambda\sigma}^{-1} \frac{\partial}{\partial x_\lambda} \qquad (6.2.25)$$

This relation will be first applied to the continuity equation (6.2.12). Consider the term

$$\sum_\sigma \frac{\partial j'_\sigma}{\partial x'_\sigma}$$

We have already seen (6.1.9) that the current density j_σ possesses the properties of a four-vector

$$j'_\sigma = \sum_{\alpha=1}^{4} L_{\sigma\alpha} j_\alpha ;$$

thus a Lorentz transformation yields the result

$$\sum_{\sigma=1}^{4} \frac{\partial j'_\sigma}{\partial x'_\sigma} = \sum_{\sigma\lambda} L_{\lambda\sigma}^{-1} \frac{\partial j'_\sigma}{\partial x_\lambda} = \sum_{\sigma\lambda\alpha} L_{\lambda\sigma}^{-1} L_{\sigma\alpha} \frac{\partial j_\alpha}{\partial x_\lambda}$$

$$= \sum_{\lambda\alpha} \delta_{\lambda\alpha} \frac{\partial j_\alpha}{\partial x_\lambda} \qquad (6.2.26)$$

$$= \sum_{\lambda=1}^{4} \frac{\partial j_\lambda}{\partial x_\lambda}$$

Since the indices σ and λ both sum over the components 1 to 4, we can replace σ by λ in our final result, and so

$$\sum_{\lambda=1}^{4} \frac{\partial j'_\lambda}{\partial x'_\lambda} = \sum_{\lambda=1}^{4} \frac{\partial j_\lambda}{\partial x_\lambda} = 0 \qquad (6.2.27)$$

Thus the equation of continuity is Lorentz invariant. For a simple transformation along the X direction the above equation could have been established just as quickly by working out all the individual terms and summing as in (6.2.16). However the equations we have given above are perfectly general and can be used for Lorentz transformations in any direction (compare §4.5).

Next let us consider the Lorentz condition (6.2.7)

$$\frac{\partial A_\lambda}{\partial x_\lambda} = 0$$

It is evident that provided A_λ transforms in the same manner as j_λ, Lorentz invariance (6.2.27) automatically follows, since both the equations for the Lorentz condition and the continuity equation have the same form

$$\sum_\lambda \frac{\partial A_\lambda}{\partial x_\lambda} = \sum_\lambda \frac{\partial j_\lambda}{\partial x_\lambda} = 0$$

The fact that A and j must transform in the same manner is easily seen in Maxwell's equations (6.2.14)

$$\sum_\lambda \frac{\partial^2 A_\mu}{\partial x_\lambda^2} = -\mu_0 j_\alpha$$

Here we shall show that

$$\sum_\lambda \frac{\partial^2}{\partial x_\lambda^2}$$

is Lorentz invariant and so A_α and j_α must transform in the same manner.

Let us therefore examine the behaviour of the double differential. Equation (6.2.25) tells us that

$$\frac{\partial}{\partial x_\sigma'} = \sum_\lambda L_{\lambda\sigma}^{-1} \frac{\partial}{\partial x_\lambda};$$

consequently

$$\sum_{\sigma=1}^{4} \frac{\partial^2}{\partial x_\sigma'^2} = \sum_{\sigma\lambda v} L_{\lambda\sigma}^{-1} L_{v\sigma}^{-1} \frac{\partial}{\partial x_\lambda} \frac{\partial}{\partial x_v} \qquad (6.2.28)$$

but from equations (6.2.20) and (6.2.23) we know that

$$L_{v\sigma}^{-1} = L_{\sigma v}$$

and

$$\sum_{\sigma=1}^{4} L_{\lambda\sigma}^{-1} L_{\sigma v} = \delta_{\lambda v}$$

Hence

$$\sum_{\sigma=1}^{4} \frac{\partial^2}{\partial x_\sigma'^2} = \sum_{\lambda v} \delta_{\lambda v} \frac{\partial}{\partial x_\lambda} \frac{\partial}{\partial x_v} = \sum_{\lambda=1}^{4} \frac{\partial^2}{\partial x_\lambda^2}$$

and since the summation over σ and λ covers the same indices

$$\sum_{\lambda=1}^{4} \frac{\partial^2}{\partial x_\lambda'^2} = \sum_{\lambda=1}^{4} \frac{\partial^2}{\partial x_\lambda^2} \qquad (6.2.29)$$

Thus the double differential remains invariant under a Lorentz transformation, and so A_α will transform in the same manner as j_α in (6.2.14)

$$\sum_{\lambda=1}^{4} \frac{\partial^2 A_\alpha}{\partial x_\lambda^2} = -\mu_0 j_\alpha \xrightarrow[\text{transformation}]{\text{Lorentz}} \sum_{\lambda=1}^{4} \frac{\partial^2 A'_\alpha}{\partial x'^2_\lambda} = -\mu_0 j'_\alpha \qquad (6.2.30)$$

A_α is therefore a four-vector which behaves like j_α and we can write its Lorentz transformation in the manner of (6.1.9)

$$A'_\lambda = \sum_{\alpha=1}^{4} L_{\lambda\alpha} A_\alpha \qquad (6.2.31)$$

6.2.3   The transformation of the fields **B** and **E**

Equation (6.2.30) shows that the Lorentz invariance of Maxwell's equations can be displayed in an explicit, and at the same time simple, manner. The transformation of the Maxwell equations in terms of the magnetic and electric fields (6.1.1) is a messy operation, but again leads to the same result—that the equations remain invariant in form under Lorentz transformations. This result is to be expected—the introduction of the four-vector A_α was made for convenience of notation and it contains no new physical notions. Nevertheless its use simplifies enormously the problem of handling relativistic problems.

In this section we shall sketch in how the fields **B** and **E** transform, and leave the details of the working out to the interested reader. Let us recall the basic definitions relating **E**, **B** and **A** (compare (6.2.2) and (6.2.3))

$$\mathbf{B} = \nabla \times \mathbf{A} \qquad \mathbf{E} = -\nabla\phi - \frac{\partial \mathbf{A}}{\partial t}$$

From the equations the x components of the fields can be written as

$$B_x = \frac{\partial A_z}{\partial y} - \frac{\partial A_y}{\partial z} \equiv B_1 = \frac{\partial A_3}{\partial x_2} - \frac{\partial A_2}{\partial x_3}$$

$$E_x = -\frac{\partial \phi}{\partial x} - \frac{\partial A_x}{\partial t} \equiv E_1 = ic \left(\frac{\partial A_4}{\partial x_1} - \frac{\partial A_1}{\partial x_4} \right) \qquad (6.2.32)$$

An inspection of the above formulations allows the y and z components to be written down immediately.

The Lorentz transformation of the components of the fields can then be carried out with the aid of equations (6.2.25) and (6.2.31)

$$\frac{\partial}{\partial x'_\sigma} = \sum_\lambda L^{-1}_{\lambda\sigma} \frac{\partial}{\partial x_\lambda} \qquad A'_\sigma = \sum_\alpha L_{\sigma\alpha} A_\alpha$$

The transformation is straightforward and merely involves reading off the matrix elements of L and L^{-1} from (6.2.18) and (6.2.22) respectively; the reader is advised to reproduce some of the results for himself. They are as follows

$$E'_x = E_x \qquad\qquad B'_x = B_x$$

$$E'_y = \gamma(E_y - vB_z) \qquad B'_\lambda = \left(B_y + \frac{v}{c^2}E_z\right)$$

(6.2.33)

$$E'_z = \gamma(E_z + vB_y) \qquad B'_z = \left(B_z - \frac{v}{c^2}E_y\right)$$

The proof of the invariance of Maxwell's equations in $\mathbf{E}$ and $\mathbf{B}$ notation is then a straightforward task.

The results in (6.2.33) can be put in a different way. Since v is in the X direction (that is, $|\mathbf{v}| = v = v_x$), all the terms involving v are components of the cross product $\mathbf{v} \times \mathbf{E}$ and $\mathbf{v} \times \mathbf{B}$; for example, with the aid of appendix 2 we find that the y component of $\mathbf{v} \times \mathbf{B}$ is

$$(\mathbf{v} \times \mathbf{B})_y = v_z B_x - v_x B_z = -vB_z$$

since $v_z = 0$. Thus we can write E_y in (6.2.33) as

$$E'_y = \gamma(\mathbf{E} + \mathbf{v} \times \mathbf{B})_y$$

and proceed in a similar manner for the other components. The results can then be summarised in a remarkably compact manner in terms of components perpendicular ($\perp$) and parallel ($\parallel$) to the X axis

$$\mathbf{E}'_\parallel = \mathbf{E}_\parallel \qquad\qquad \mathbf{B}'_\parallel = \mathbf{B}_\parallel$$

(6.2.34)

$$\mathbf{E}'_\perp = \gamma(\mathbf{E} + \mathbf{v} \times \mathbf{B})_\perp \qquad \mathbf{B}'_\perp = \gamma\left(\mathbf{B} - \frac{\mathbf{v}}{c^2} \times \mathbf{E}\right)_\perp$$

The above equations do not contain any specific reference to the X, Y and Z axes. Since $\parallel$ refers to a Lorentz transformation along the X axis with velocity v, equations (6.2.34) can be used for transformation in any direction with the subscripts $\parallel$ and $\perp$ referring to the direction of $\mathbf{v}$.

Equations (6.2.34) show that what one calls an electric field or a magnetic field depends on one's reference frame. Supposing we have an observer at rest relative to some physical apparatus and that he is equipped with apparatus to detect electric and magnetic fields. Let his equipment record the results

$$\mathbf{E} = \mathbf{E}_0 \qquad \mathbf{B} = 0$$

that is, the presence of an electric field only. A second observer, similarly equipped, but moving past the apparatus with velocity $\mathbf{v}$ then records

$$\mathbf{E}'_\parallel = \mathbf{E}_{0\parallel} \qquad \mathbf{B}'_\parallel = 0$$

$$\mathbf{E}'_\perp = \gamma\mathbf{E}_{0\perp} \qquad \mathbf{B}'_\perp = -\frac{\gamma}{c^2}(\mathbf{v} \times \mathbf{E}_0)_\perp$$

that is an electric field and also a magnetic field at right angles to the direction of motion. Thus electric and magnetic fields are different aspects of the same electromagnetic phenomena, and how one describes them depends on the state of relative motion of the observer and the electromagnetic system.

In closing this section we note a useful shorthand which can be associated with the field components (6.2.32). If we introduce an antisymmetric tensor of the second rank (§4.9)

$$F_{\alpha\rho} = \frac{\partial A_\beta}{\partial x_\alpha} - \frac{\partial A_\alpha}{\partial x_\beta} = -F_{\beta\alpha} \tag{6.2.35}$$

then following the pattern of (6.2.32) it is not difficult to show that the components of $F_{\alpha\beta}$ are

$$F_{\alpha\beta} = \begin{bmatrix} 0 & B_3 & -B_2 & \dfrac{-iE_1}{c} \\ -B_3 & 0 & B_1 & \dfrac{-iE_2}{c} \\ B_2 & -B_1 & 0 & \dfrac{-iE_3}{c} \\ \dfrac{iE_1}{c} & \dfrac{iE_2}{c} & \dfrac{iE_3}{c} & 0 \end{bmatrix} \tag{6.2.36}$$

6.3 Solutions of Maxwell's equations

6.3.1 The effects of Galilean and Lorentz transformations upon propagation of electromagnetic waves

Consider the propagation of electromagnetic waves in a charge-free space. In this condition the four current density $j_\alpha = 0$ for all components α, and Maxwell's equations in four-vector notation (6.2.11) then become

$$\sum_{\lambda=1}^{4} \frac{\partial^2 A_\alpha}{\partial x_\lambda^2} = \nabla^2 A_\alpha - \frac{1}{c^2} \frac{\partial^2 A_\alpha}{\partial t^2} = 0 \tag{6.3.1}$$

Let us assume that the electromagnetic waves move in the X direction. The wave nature of the motion can be represented by an equation of the form

$$A_\alpha = \hat{A}_\alpha \exp\left[i(2\pi/\lambda)(x - ut)\right] \tag{6.3.2}$$

according to the usual notions of wave theory. The choice of $x - ut$ rather $x + ut$ ensures propagation along the $+X$ direction. In the above equation λ represents the wavelength of the wave, u its velocity of propagation and $\hat{A}$ the non-oscillatory part of the wave ('the amplitude').

The insertion of equation (6.3.2) into (6.3.1) yields the result

$$(1 - u^2/c^2) \hat{A}_\alpha = 0 \tag{6.3.3}$$

which requires $u = c$. Thus the term c which essentially appears in Maxwell's equations (6.1.1) as a constant of proportionality represents the velocity of propagation of the electromagnetic waves. Furthermore if we make a Lorentz transformation so that A_α becomes

$$A'_\alpha = \hat{A}'_\alpha \exp[i(2\pi/\lambda')(x'-u't')] \tag{6.3.4}$$

then the form of equation (6.3.1) remains unchanged (compare (6.2.30)

$$\sum_{\lambda=1}^{4} \frac{\partial^2 A'_\alpha}{\partial x'^2_\lambda} = \sum_{\lambda=1}^{4} \frac{\partial^2 A_\alpha}{\partial x^2_\lambda} = 0$$

and the analogous equation to (6.3.3)

$$(1-u'^2/c^2)\,\hat{A}'_\alpha = 0$$

leads to the requirement $u=u'=c$. Thus the requirement of Lorentz invariance for physical equations has led us back to Einstein's original postulate (§2.2) of the invariance of the velocity of light in all reference frames which move uniformly with respect to each other.

Next let us compare the effects of a Lorentz and Galilean transformation. In terms of the ether postulate of chapter 1 (§1.3) we shall assume that Maxwell's equations hold in the ether reference frame. We shall therefore again start with equations (6.3.1) and (6.3.2)

$$\left(\frac{\partial^2}{\partial x^2} + \frac{\partial^2}{\partial y^2} + \frac{\partial^2}{\partial z^2} - \frac{1}{c^2}\frac{\partial}{\partial t^2}\right) A_\alpha = 0$$

$$A_\alpha = \hat{A}_\alpha \exp[i(2\pi/\lambda)(x-ut)]$$

where the second equation represents the wave propagating in the X direction. These equations lead to the result $u=c$ as we have shown above. In what follows we shall concentrate on the oscillating part of A_α; the term $2\pi/\lambda$ is not necessary for our considerations and will be dropped. We then have

$$\exp[i(2\pi/\lambda)(x-ut)] \equiv \exp[i(x-ct)] \tag{6.3.5}$$

A Galilean transformation (1.1.2) in the X direction tells us that

$$x' = x - vt$$
$$y' = y$$
$$z' = z$$
$$t' = t$$

where v is the velocity of the transformed reference frame. The oscillatory term in (6.3.5) then becomes

$$\begin{aligned}\exp[i(x-ct)] &\rightarrow \exp[i(x'+vt'-ct')]\\ &= \exp\{i[x'-(c-v)t']\}\\ &\equiv \exp[i(x'-u't')]\end{aligned} \tag{6.3.6}$$

Thus we find that the velocity of propagation of the electromagnetic waves (that is, light) in the transformed system is

$$u' = c - v$$

as already anticipated in equation (1.3.1). Thus the Galilean transformation fails to maintain the principle of the invariance of the velocity of light in all reference frames. If we make the same examination for a Lorentz transformation (2.3.15) we find

$$
\begin{aligned}
\exp i(x - ct) &\rightarrow \exp\{i\gamma\,[x' + vt' - c(t' + vx'/c^2)]\} \\
&= \exp\{i\gamma\,[x'(1 - v/c) - (c - v)\,t']\} \\
&= \exp\{i\gamma\,(1 - v/c)\,[x' - (c - v)\,t'/(1 - v/c)]\} \quad (6.3.7) \\
&= \exp[i\gamma\,(1 - v/c)\,(x' - ct')] \\
&\equiv \exp[i(x' - u't')]
\end{aligned}
$$

and we see that $u' = c$. Thus, as we have shown before, the velocity of light remains invariant under Lorentz transformations.

We have previously shown (6.2.29) that the second-order differential

$$\sum_{\lambda=1}^{4} \frac{\partial^2}{\partial x_\lambda^2}$$

remains invariant under a Lorentz transformation. If the Galilean transformation (1.1.2) is used in conjunction with the rules for partial differentiation (6.2.15) we find that

$$\sum_{\lambda=1}^{4} \frac{\partial}{\partial x_\lambda^2} \nrightarrow \sum_{\lambda=1}^{4} \frac{\partial^2}{\partial x_\lambda'^2}$$

Thus the equation (6.3.1)

$$\sum_{\lambda=1}^{4} \frac{\partial^2 A_\alpha}{\partial x_\lambda^2} = 0$$

does not remain invariant in form under a Galilean transformation, and furthermore the velocity of light appears to change.

6.3.2 · Relativistic wave equations

The oscillatory part of equation (6.3.2) was written for a wave travelling in the $+X$ direction. This equation is often given a more manifestly covariant form. A wave moving in an arbitrary direction, that is, not necessarily along the X axis can be represented by a wave vector $\mathbf{k}$

$$\mathbf{k} = (2\pi/\lambda)\,\mathbf{l} \qquad (6.3.8)$$

where $\mathbf{l}$ is a unit vector which points along the direction of propagation of the wave. Thus the term $2\pi/\lambda$ in (6.3.2)

$$A_\alpha = \hat{A}_\alpha \exp[i(2\pi/\lambda)\,(x - ut)]$$

can be written as

$$(2\pi/\lambda)\,\mathbf{l}\cdot\mathbf{x}=\mathbf{k}\cdot\mathbf{x} \tag{6.3.9}$$

At the same time we can replace u by c (6.3.3) and $2\pi u/\lambda$ becomes

$$2\pi c/\lambda=2\pi v=\omega \tag{6.3.10}$$

where v and ω represent frequency and angular velocity respectively. Thus equation (6.3.2) can be converted to

$$A_\alpha=\hat{A}_\alpha\exp[i(\mathbf{k}\cdot\mathbf{x}-\omega t)] \tag{6.3.11}$$

for a wave travelling in the direction of $\mathbf{k}$. From this point it is a simple step to write

$$k_4=i\omega/c \qquad x_4=ict$$

and the exponent becomes

$$\mathbf{k}\cdot\mathbf{x}-\omega t=\mathbf{k}\cdot\mathbf{x}+k_4 x_4=\sum_{\sigma=1}^{4}k_\sigma x_\sigma$$

Equation (6.3.1) then assumes the form

$$A_\alpha=\hat{A}_\alpha\exp\left(i\sum_\sigma k_\sigma x_\sigma\right)=\hat{A}_\alpha\exp(ikx) \tag{6.3.12}$$

In the final form we have dropped the summation over σ, $\sum_\sigma$, since we shall have occasion to use this equation repeatedly in the rest of this section. Nevertheless, it must be understood that whenever the term $\exp(ikx)$ appears then the summation over σ is implied.

Thus the exponent in (6.3.12) now displays a Lorentz invariant form.* It is instructive to write this equation in the language of wave mechanics. If we use the de Broglie and Planck relations, which connect wavelength with momentum and frequency with energy, we have

$$\lambda=\frac{h}{p}\longrightarrow\mathbf{p}=\frac{h}{\lambda}\mathbf{l}; \qquad E=hv$$

where the first equation represents the simple one-dimensional de Broglie relation. If these equations are inserted into the exponent of (6.3.11) we find with the aid of equations (6.3.4) and (6.3.10) that

$$\begin{aligned}i(\mathbf{k}\cdot\mathbf{x}-\omega t)&=i\,2\pi[(\mathbf{l}/\lambda)\cdot\mathbf{x}-vt)]\\&=i(2\pi/h)(\mathbf{p}\cdot\mathbf{x}-Et)\end{aligned} \tag{6.3.13}$$

---

* Although we have given the exponent a Lorentz invariant *form*, we have not actually demonstrated Lorentz invariance. This problem is left as an exercise to the reader (a transformation in the X direction is adequate).

It is customary to use the symbol $\hbar$ for $h/2\pi$; thus we can write in the manner of (6.3.11) and (6.3.12)

$$A_\alpha = \hat{A}_\alpha \exp[(i/\hbar)(\mathbf{p}\cdot\mathbf{x} - Et)] = \hat{A}_\alpha \exp(ipx/\hbar) \qquad (6.3.14)$$

where p denotes the momentum four-vector and once again we have dropped the summation over the four-components of p and x.

If we apply Maxwell's equation (6.3.1) to (6.3.14) we find

$$\sum_\lambda \frac{\partial^2 A_\alpha}{\partial x_\lambda^2} = -\frac{1}{\hbar^2}(p_x^2 + p_y^2 + p_z^2 - E^2/c^2)\,A_\alpha$$

$$= -\frac{1}{\hbar^2}(\mathbf{p}^2 - E^2/c^2)\,A_\alpha$$

$$= 0$$

Thus we find that

$$E^2 = \mathbf{p}^2 c^2$$

which is the Einstein equation (4.3.2) for massless particles. Therefore free electromagnetic waves have the same mechanical properties as massless particles, that is, photons, when the quantisation conditions are applied to them.

The Maxwell equations (6.3.1) and (6.2.7)

$$\sum_\lambda \frac{\partial^2 A_\alpha}{\partial x_\lambda^2} = \nabla^2 A_\alpha - \frac{1}{c^2}\frac{\partial^2 A_\alpha}{\partial t^2} = 0$$

$$\sum_\alpha \frac{\partial A_\alpha}{\partial x_\alpha} = 0$$

are in fact a specialised case of a more generalised set of wave equations in relativistic quantum mechanics

$$\nabla^2 A_\alpha - \frac{1}{c^2}\frac{\partial^2 A_\alpha}{\partial t^2} - \frac{m_0^2 c^2}{\hbar^2}A_\alpha = 0$$

$$\sum_\alpha \frac{\partial A_\alpha}{\partial x_\alpha} = 0 \qquad (6.3.15)$$

The first of these equations leads to the Einstein equation (4.3.2) for particles of rest mass m_0

$$E^2 = \mathbf{p}^2 c^2 + m_0^2 c^4$$

whilst the second leads to the requirement that the particles have spin one (we shall discuss the implications of the second equation for photons in §6.3.3 and 6.3.4). At least nine particles with spin one and finite rest mass are known to exist in elementary particle physics. For readers familiar with the SU_3 classification in elementary particle physics, the particles are the various charged states of the nonet ϱ, ω, ϕ, K^*. Only one (the photon) is known with zero rest mass.

The first of equations (6.3.15)

$$\left(\nabla^2 - \frac{1}{c^2}\frac{\partial^2}{\partial t^2} - \frac{m_0^{\nu}c^2}{\hbar^2}\right) A_\alpha = 0$$

in slightly different forms has a quite general validity in relativistic quantum mechanics and field theory.[†] For example if A_α has only one component so that it is no longer associated with space and time components, the resultant wave equation (the Klein–Gordon equation) is suitable for describing particles with rest mass m_0 and spin zero. Again at least nine particles are known with this property—the pions, which are mentioned in chapter 3, belong to this group. The wave equation for particles with spin one half (electrons, muons, protons and neutrinos) is somewhat more complicated but the associated function A (usually denoted by the symbol ψ) also obeys the first of equations (6.3.15).

6.3.3 Solutions in the Coulomb gauge

The solution (6.3.12)

$$A_\alpha = \hat{A}_\alpha \exp(ikx)$$

in conjunction with the Lorentz gauge condition (6.2.7)

$$\sum_\alpha \frac{\partial A_\alpha}{\partial x_\alpha} = 0$$

implies that we can write

$$\sum_\alpha k_\alpha \hat{A}_\alpha = \mathbf{k} \cdot \hat{\mathbf{A}} + k_4 \hat{A}_4 = 0 \tag{6.3.16}$$

We emphasised in §6.2.1 that A_α is arbitrary in its definition and was introduced as an auxiliary function for mathematical convenience. We shall again exploit this arbitrariness by making $\hat{A}_4 = 0$, then equation (6.3.14) becomes

$$\sum_\alpha k_\alpha \hat{A}_\alpha = \mathbf{k} \cdot \hat{\mathbf{A}} = 0 \tag{6.3.17}$$

Thus $\hat{\mathbf{A}}$ is then perpendicular to $\mathbf{k}$.

The choice of $\hat{A}_4 \equiv A_4 = 0$ is known as the *Coulomb gauge condition*. The use of $A_4 = 0$ for a charge-free system (that is, $j_\alpha = 0$ in (6.2.14)) makes sense, since the scalar potential $\phi = cA_4/i$ (6.2.6) arises from the presence of the charge in electrostatics.

Despite the apparent mathematical juggling, which we have performed above, everything is still consistent with Maxwell's equations, and in particu-

---

[†] The interested reader can pursue the matter further in the present author's *Notes on Elementary Particle Physics* and *The Physics of Elementary Particles* published by ⁻ergamon ⁻ress.

lar with the electric and magnetic fields **E** and **B**. As we have emphasised before, it is these fields that are measured in experiments, and not A which is a convenient mathematical tool.

Let us check that everything is still consistent with Maxwell's equations. In equations (6.2.2) and (6.2.3) we defined **A** and ϕ in terms of **B** and **E** respectively

$$\mathbf{B} = \nabla \times \mathbf{A} \qquad \mathbf{E} = -\nabla\phi - \frac{\partial \mathbf{A}}{\partial t}$$

Now $\phi = cA_4/i$ (6.2.6), and so if we insert (6.3.12) into the above equations and employ the condition (6.3.17) we find

$$\mathbf{B} = i\mathbf{k} \times \hat{\mathbf{A}} \exp(ikx) \qquad \mathbf{E} = i\omega\hat{\mathbf{A}} \exp(ikx) \qquad (6.3.18)$$

But according to Maxwell's equations (6.1.1) we expect

$$\nabla \cdot \mathbf{B} = 0 \qquad \nabla \cdot \mathbf{E} = 0$$

for a charge-free electromagnetic system. The operator div acting on **B** and **E** in (6.3.18) yields

$$\nabla \cdot \mathbf{B} = -\mathbf{k} \cdot (\mathbf{k} \times \hat{\mathbf{A}}) \exp(ikx) = 0$$
$$\nabla \cdot \mathbf{E} = -\mathbf{k} \cdot \hat{\mathbf{A}}\omega \exp(ikx) = 0$$

and so Maxwell's equation are satisfied. The first result follows from the fact that the vector product $(\mathbf{k} \times \hat{\mathbf{A}})$ is perpendicular to **k**, whilst the second result arises from (6.3.17), $\mathbf{k} \cdot \hat{\mathbf{A}} = 0$.

Equations (6.3.18) reveal an interesting relation between **k**, **E** and **B** for free electromagnetic waves. Equation (6.3.18) shows that **E** is perpendicular to **k**

$$\mathbf{k} \cdot \mathbf{E} = i\omega\mathbf{k} \cdot \hat{\mathbf{A}} \exp(ikx) = 0 \qquad (6.3.19)$$

by virtue of (6.3.17). Alternatively, given an oscillating free electromagnetic wave, the Maxwell condition $\nabla \cdot \mathbf{E} = 0$, automatically implies that **k** and **E** are perpendicular. Next consider the behaviour of **B**; if we recall our definition of **k** in (6.3.8) and also use equation (6.3.10), we find that

$$\mathbf{k} = \frac{2\pi}{\lambda}\mathbf{l} = 2\pi\frac{v}{c}\mathbf{l} = \frac{\omega}{c}\mathbf{l} \qquad (6.3.20)$$

where **l** is a unit vector pointing along the direction of propagation of the wave; then from (6.3.18)

$$\mathbf{B} = \frac{\mathbf{k}}{\omega} \times \mathbf{E} = \frac{\mathbf{l}}{c} \times \mathbf{E} \qquad (6.3.21)$$

Thus **B** is at right angles to both **E** and **k**, and at the same time it is related to **E** by a constant of proportionality $1/c$.

The mutual orthogonality of **k**, **E** and **B** can be displayed as shown in figure 6.1. In this figure we have also introduced unit vectors $\mathbf{n}_1$ and $\mathbf{n}_2$,

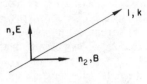

Figure 6.1

defined as

$$\mathbf{E}=\mathbf{n}_1 E_0 \qquad \mathbf{B}=\mathbf{n}_2 \frac{E_0}{c} \qquad (6.3.22)$$

where

$$E_0 = i\omega \hat{A}_0 \exp(ikx) = i\omega \hat{A}_0 \exp[i(\mathbf{k}\cdot\mathbf{x}-\omega t)]$$

Thus $\mathbf{n}_1$ and $\mathbf{n}_2$ lie in the plane perpendicular to $\mathbf{k}$, and we find from (6.3.21) that

$$\mathbf{n}_2 = \mathbf{l}\times\mathbf{n}_1 \qquad \mathbf{n}_1 = -\mathbf{l}\times\mathbf{n}_2 \qquad (6.3.23)$$

(compare appendix 2).

6.3.4   The 'spin' of a photon [*]

In this section the helicity (§5.4) of the photon will be examined. In chapter 5 it was emphasised that the understanding of angular momentum at the atomic level could only be properly understood within the framework of quantum mechanics, and, as we shall see below, we shall have to adopt some of the notions of this subject.

Let us introduce a new complex vector

$$\mathbf{\psi}_+ = \mathbf{E}+ic\mathbf{B} \qquad \mathbf{\psi}_- = \mathbf{E}-ic\mathbf{B} \qquad (6.3.24)$$

The physical significance of these combinations is readily grasped from (6.3.22)

$$\mathbf{E}\pm ic\mathbf{B} = (\mathbf{n}_1\pm i\mathbf{n}_2)\,E_0 = i\omega\hat{A}_0\,(\mathbf{n}_1\pm i\mathbf{n}_2)\exp(ikx)$$

Now

$$i = \exp(i\pi/2) \qquad ; \qquad -i = \exp(i\pi/2)$$

so that

$$\mathbf{E}\pm ic\mathbf{B} = i\omega A_0\,[\mathbf{n}_1\,\exp(ikx)+\mathbf{n}_2\,\exp\{i(kx\pm\pi/2)\}]$$

Thus $\mathbf{\psi}_+$ and $\mathbf{\psi}_-$ represent situations where the magnetic wave is $\pi/2$ and

---

[*] The method of R. H. Good (*Phys. Rev.* **105**, 1914, 1957) is followed.

$-\pi/2$ respectively out of phase with respect to the electric wave. Classically this situation can be explored in terms of the rotations of the electric and magnetic field vectors about the direction of propagation of the waves (an excellent discussion may be found in Jackson[†]). We wish to examine the implications of equations (6.2.25), however, from the viewpoint of quantum mechanics.

First, however, let us return to Maxwell's equations for free electromagnetic waves (that is, $j_x = 0$), then from (6.1.1) we can write

$$\mathbf{V} \times \mathbf{B} = \frac{1}{c^2}\frac{\partial \mathbf{E}}{\partial t} \qquad \mathbf{V} \times \mathbf{E} = -\frac{\partial \mathbf{B}}{\partial t} \qquad (6.3.25)$$
$$\mathbf{V} \cdot \mathbf{B} = 0 \qquad \mathbf{V} \cdot \mathbf{E} = 0$$

The linear combinations ψ_+ and ψ_- are divergenceless like $\mathbf{E}$ and $\mathbf{B}$ in the above equations, since

$$\mathbf{V} \cdot \psi_+ = \mathbf{V} \cdot (\mathbf{E} + ic\mathbf{B}) = \mathbf{V} \cdot \mathbf{E} + ic\mathbf{V} \cdot \mathbf{B} = 0 \qquad (6.3.26)$$

and similarly

$$\mathbf{V} \cdot \psi_- = 0$$

Now let us examine the top pair of equations of (6.3.25); if we take the combination

$$\mathbf{V} \times (\mathbf{E} + ic\mathbf{B}) = -\frac{\partial \mathbf{B}}{\partial t} + \frac{i}{c}\frac{\partial \mathbf{E}}{\partial t}$$
$$= \frac{i}{c}\frac{\partial}{\partial t}(\mathbf{E} + ic\mathbf{B})$$

on the other hand

$$\mathbf{V} \times (\mathbf{E} - ic\mathbf{B}) = -\frac{i}{c}\frac{\partial}{\partial t}(\mathbf{E} - ic\mathbf{B})$$

Thus we have found that

$$\mathbf{V} \times \psi_+ = \frac{i}{c}\frac{\partial \psi_+}{\partial t} \qquad \mathbf{V} \times \psi_- = -\frac{i}{c}\frac{\partial \psi_-}{\partial t} \qquad (6.3.27)$$

We can express the above results in the notation of (6.3.22). If we write

$$\psi_+ = (\mathbf{E} + ic\mathbf{B}) = (\mathbf{n}_1 + i\mathbf{n}_2) E_0$$

then the first of equations (6.3.27) gives

$$i\mathbf{k} \times (\mathbf{n}_1 + i\mathbf{n}_2) E_0 = -\frac{i}{c} \cdot i\omega (\mathbf{n}_1 + i\mathbf{n}_2) E_0$$

[†] JACKSON, J.D., *Classical Electrodynamics*, Wiley 1962.

or

$$i\mathbf{k} \times (\mathbf{n}_1 + i\mathbf{n}_2) = \frac{\omega}{c}(\mathbf{n}_1 + i\mathbf{n}_2) \tag{6.3.28}$$

The validity of the above equation is easily checked from (6.3.23). At the same time the second of equations (6.3.27) gives

$$i\mathbf{k} \times (\mathbf{n}_1 - i\mathbf{n}_2) = -\frac{\omega}{c}(\mathbf{n}_1 - i\mathbf{n}_2) \tag{6.3.29}$$

At this stage the ideas of quantum theory must be introduced by injecting the notation of equations (6.3.13)

$$\exp(ikx) \to \exp(ipx/\hbar) \to \exp[i(\mathbf{p}\ \mathbf{x} - Et)/\hbar]$$

similarly

$$\mathbf{k} \to \mathbf{p} \qquad \omega \to E = c|\mathbf{p}|$$

where the last relation follows from the massless nature of the photon (4.3.2). Thus equations (6.3.28) and (6.3.29) become

$$\frac{i\mathbf{p}}{|\mathbf{p}|} \times (\mathbf{n}_1 + i\mathbf{n}_2) = (\mathbf{n}_1 + i\mathbf{n}_2)$$
$$\frac{i\mathbf{p}}{|\mathbf{p}|} \times (\mathbf{n}_1 - i\mathbf{n}_2) = -(\mathbf{n}_1 - i\mathbf{n}_2) \tag{6.3.30}$$

In going from classical electromagnetic theory to quantum theory we must simultaneously change our description of the electromagnetic system from a beam of waves to a beam of photons ('the wave–particle duality'). However, in order to avoid getting too deeply involved in quantum mechanics we shall not pursue the problems of the transition any further.

Let us go back to equations (6.3.30). The x, y, z components of these equations can also be expressed in terms of the asymmetric tensor ε_{jkl} (compare (5.2.5) to (5.2.7)), where $j, k, l = 1, 2, 3$. Thus the first of equations (6.3.30) can be written as

$$i \sum_{kl} \varepsilon_{jkl} \frac{p_k}{|\mathbf{p}|}(\mathbf{n}_1 + i\mathbf{n}_2)_l = (\mathbf{n}_1 + i\mathbf{n}_2)_j \tag{6.3.31}$$

If we now define matrices S_k with elements*

$$(S_k)_{jl} = i\varepsilon_{jkl} \tag{6.3.32}$$

equation (6.3.31) becomes

$$\sum_{kl} (S_k)_{jl} \frac{p_k}{|\mathbf{p}|}(\mathbf{n}_1 + i\mathbf{n}_2)_l = (\mathbf{n}_1 + i\mathbf{n}_2)_j$$

* The matrices are displayed in equations (6.3.35).

or equivalently

$$\sum_{l}\left(\frac{\mathbf{S}\cdot\mathbf{p}}{|\mathbf{p}|}\right)_{jl}(\mathbf{n}_1+i\mathbf{n}_2)_l=(\mathbf{n}_1+i\mathbf{n}_2)_j \qquad (6.3.33)$$

Similarly the second of equations (6.3.30) yields

$$\sum_{l}\left(\frac{\mathbf{S}\ \mathbf{p}}{|\mathbf{p}|}\right)_{jl}(\mathbf{n}_1-i\mathbf{n}_2)_l=-(\mathbf{n}_1-i\mathbf{n}_2)_j \qquad (6.3.34)$$

Now we have seen equations like these before; the expression $\mathbf{S}\cdot\mathbf{p}/|\mathbf{p}|$ is the helicity term of §5.5, but as we are now talking in the language of quantum mechanics $\mathbf{S}\cdot\mathbf{p}/|\mathbf{p}|$ is the helicity operator, whilst $(\mathbf{n}_1\pm i\mathbf{n}_2)$ are the helicity states with eigenvalues ± 1. These values lead us to suspect that the 'spin' of the photon is one—this suspicion will be verified below.

Equation (6.3.32) implies that the components of $\mathbf{S}$ read

$$S_1=\begin{bmatrix} 0 & 0 & 0 \\ 0 & 0 & -i \\ 0 & i & 0 \end{bmatrix} \quad S_2=\begin{bmatrix} 0 & 0 & i \\ 0 & 0 & 0 \\ -i & 0 & 0 \end{bmatrix} \quad S_3=\begin{bmatrix} 0 & -i & 0 \\ i & 0 & 0 \\ 0 & 0 & 0 \end{bmatrix} \qquad (6.3.35)$$

We find immediately that

$$\mathbf{S}^2=\begin{bmatrix} 2 & 0 & 0 \\ 0 & 2 & 0 \\ 0 & 0 & 2 \end{bmatrix}=1(1+1)\begin{bmatrix} 1 & 0 & 0 \\ 0 & 1 & 0 \\ 0 & 0 & 1 \end{bmatrix} \qquad (6.3.36)$$

Now the eigenvalues of the spin operator $\mathbf{S}^2$ in quantum mechanics are $s(s+1)$, where s is the spin quantum number of the eigenstates (compare (5.1.2)); thus we may conclude that the spin of the photon is one.

The association of the matrices (6.3.35) to the equations (6.3.33) and (6.3.34) is slightly complicated as $\mathbf{n}_1$ and $\mathbf{n}_2$ are linked with the direction of $\mathbf{p}$, that is $\mathbf{k}$ (compare figure 6.1). For simplicity let us assume that $\mathbf{p}$ is the direction of the 3 (that is the Z) axis in configuration space so that

$$p_1=p_2=0 \qquad p_3=p$$

then

$$\frac{\mathbf{S}\cdot\mathbf{p}}{|\mathbf{p}|}=\begin{bmatrix} 0 & -i & 0 \\ i & 0 & 0 \\ 0 & 0 & 0 \end{bmatrix} \qquad (6.3.37)$$

and equations (6.3.33) and (6.3.34) are immediately satisfied if we write $\mathbf{n}_1$ and $\mathbf{n}_2$ as

$$\mathbf{n}_1=\begin{bmatrix} 1 \\ 0 \\ 0 \end{bmatrix} \qquad \mathbf{n}_2=\begin{bmatrix} 0 \\ 1 \\ 0 \end{bmatrix} \qquad (6.3.38)$$

Solutions for arbitrary directions are complicated and will not be pursued further.

6.4 Invariants of the electromagnetic system

We have seen in §6.1.2 that the total charge remains unaltered during Lorentz transformations and in §6.2.2 that Maxwell's equations remain invariant in form. Up to now, however, we have not mentioned anything about possible Lorentz invariants for the fields.

Since both A and j are four-vectors it is apparent from our previous considerations that

$$\sum_{\alpha=1}^{4} A_\alpha A_\alpha \quad \text{and} \quad \sum_{\alpha=1}^{4} j_\alpha j_\alpha$$

are Lorentz invariants. However, A_α was introduced as a mathematical convenience and a current in isolation (that is, in the absence of electromagnetic fields) has no great physical significance. Thus both the invariants given above possess no particular interest. A different story, however, is the interaction of the current with the electromagnetic field. The action of a field (electric and/or magnetic) on a current is a physical process which is capable of measurement. The simplest term we can then combine with j_α to yield a Lorentz invariant is A_α

$$A = \sum_{\alpha=1}^{4} j_\alpha A_\alpha \tag{6.4.1}$$

where the A on the left is used to represent a probability amplitude for the interaction.

If properly handled, this seemingly simple relation contains the whole of classical electrodynamics! If appropriate quantum mechanical conditions are introduced, it also covers the whole of quantum electrodynamics, both at the level of atomic and elementary particle physics. In recent years attempts have been made to test the possibility of deviations from the basic laws of quantum electrodynamics but none have been found (we shall return to this point in §6.6).

6.5 The Lorentz force

6.5.1 Derivation

The terms $\sum_\alpha A_\alpha A_\alpha$, $\sum_\alpha j_\alpha j_\alpha$ and $\sum_\alpha j_\alpha A_\alpha$ of §6.4 are Lorentz scalars; that is, they remain unaltered in value under Lorentz transformations. In this section we shall exploit both the property of Lorentz invariance and the transformation properties of four-vectors to develop a physical equation—the Lorentz force on a charged particle.

Consider a charge Q at rest in an electric field of strength $\mathbf{E}$; the elementary laws of electrostatics tell us that the force $\mathbf{f}$ experienced by the charge is

$$\mathbf{f}_R = \left(m_0 \frac{du}{dt} \right)_R = Q \mathbf{E}_R \tag{6.5.1}$$

where we shall show the subscript R to designate the rest frame.

Now supposing we wish to construct a more general relation to cover the situation when the charge is not necessarily at rest. One way of doing this would be to perform a Lorentz transformation, using the knowledge we have already gained in previous sections. Another way, which will be developed below, relies upon the exploitation of the general properties of Lorentz invariance and four-vectors.

Now equation (6.5.1) contains the force **f** on its left-hand side. It was stated in § 4.9 that force is not a four-vector. However, the four-force F_λ (4.9.3)

$$F_\lambda = \frac{dP_\lambda}{d\tau} = \gamma_u \frac{dP_\lambda}{dt}$$

is related to **f** as we show below. In the above equation P_λ is the momentum four-vector, and so in the notation of (4.2.7) and (4.2.8)

$$F = \left(\gamma_u \frac{d\mathbf{p}}{dt}, \frac{i}{c} \gamma_u \frac{dE}{dt} \right)$$

But $d\mathbf{p}/dt$ is the standard definition of force **f** (compare (4.9.2)), also from (4.3.2) and (4.2.9)

$$
\begin{aligned}
\frac{dE}{dt} &= \frac{d}{dt} \sqrt{(\mathbf{p}^2 c^2 + m_0^2 c^4)} \\
&= \frac{d}{dt} \sqrt{[(p_x^2 + p_y^2 + p_z^2) c^2 + m_0^2 c^4]} \\
&= \frac{1}{E} \left(p_x \frac{dp_x}{dt} + p_y \frac{dp_y}{dt} + p_z \frac{dp_z}{dt} \right) c^2 \\
&= \mathbf{u} \cdot \mathbf{f}
\end{aligned}
\tag{6.5.2}
$$

Thus the components of F are

$$F = \left(\gamma_u \mathbf{f}, \frac{i}{c} \gamma_u \mathbf{u} \cdot \mathbf{f} \right) \tag{6.5.3}$$

and it is apparent that in the rest frame $(\mathbf{u} = 0)$ we obtain the four-vector

$$F_R = (\mathbf{f}_R, 0) \tag{6.5.4}$$

That is, the spatial components coincide with the nonrelativistic concept of force.

Let us now return to equation (6.5.1) for the charge at rest

$$\mathbf{f}_R = Q \mathbf{E}_R$$

If we consider the x, or rather 1, component, then from (6.2.36)

$$f_{1R} = Q E_{1R} = Q i c F_{14R}$$

and from (6.5.4)

$$f_{1R}=F_{1R}=QicF_{14R} \qquad (6.5.5)$$

Now in the above equation Q is a Lorentz invariant (§6.1.2) and so is a constant for all practical purposes; on the other hand F_{1R} is the first component of a four-vector and will undergo Lorentz transformations as such, therefore the right-hand side must also behave like the first component of a four-vector. Upon first inspection, the right-hand side does not appear to possess any such property, since it is composed of F_{14}, that is, the first and fourth components of the four-vectors A_α and x_β (6.2.35). Consider, however, the term ic, which at first appears to be yet another constant. In fact ic is the fourth component of the four-velocity U_λ in its rest frame, since from (4.2.3)

$$U=(\gamma_u\mathbf{u}, i\gamma_u c)$$

and for $\mathbf{u}=0$ then $\gamma_u=1$ and so

$$U_\lambda=0, 0, 0, ic \qquad (6.5.6)$$

Thus we can rewrite equation (6.5.5) as

$$\begin{aligned}F_{1R}&=QicF_{14R}\\&=Q(F_{14}U_4)_R\\&=Q\sum_{\beta=1}^{4}(F_{1\beta}U_\beta)_R\end{aligned} \qquad (6.5.7)$$

We have shown in §6.2.2 that the occurrence of repeated indices leads to Lorentz invariant quantities (the summation is sometimes called the saturation of the indices). Thus the right-hand side of (6.5.7) contains only the first component of a four-vector (for purposes of Lorentz transformations) and so it will behave in the same way as the left-hand side.

From equation (6.5.7) it is a simple step to generalise to all four components and all reference frames

$$F_\alpha=m_0\frac{dU_\alpha}{d\tau}=Q\sum_{\beta=1}^{4}F_{\alpha\beta}U_\beta \qquad (6.5.8)$$

where the second term is constructed from the definition of four-acceleration (4.9.3).

Equation (6.4.9) is the equation of the Lorentz force in four-vector notation. Prior to the work of Einstein, Lorentz had introduced the equation as an axiom, whereas it is a consequence of the relativity principle. The equation can be cast into a somewhat more familiar form by considering, say, the first component in conjunction with (6.5.3) and (6.2.36)

$$\begin{aligned}F_1=\gamma_u f_1&=Q(F_{11}U_1+F_{12}U_2+F_{13}U_3+F_{14}U_4)\\&=Q\gamma_u\left(B_3u_2-B_2u_3-\frac{i}{c}E_1 ic\right)\\&=Q\gamma_u(\mathbf{u}\times\mathbf{B}+\mathbf{E})_1\end{aligned}$$

Thus we have obtained the $x(\equiv 1)$ component of the normal equation for the Lorentz force, and in general

$$\mathbf{f} = Q(\mathbf{u} \times \mathbf{B} + \mathbf{E}) \tag{6.5.9}$$

It is evident that in the limit $\mathbf{u} \to 0$ we return to our original electrostatic equation (6.5.1).

The Lorentz force equation has many applications (we shall discuss some in the next section). However, for our present considerations the importance of the equation lies in its method of derivation. In dimensional analysis it is possible to construct equations from the appropriate physical variables by getting the dimensions to balance on both sides of the equation. Here we have introduced the additional requirement that *both sides of the equation behave in the same way under Lorentz transformations; that is, the equation is covariant in form.* The principle is widely used in constructing equations in elementary particle physics and can be extended to other transformations; for example it is no use expecting an equation to be independent of displacements if one side is dependent on the position variable x.

Whilst equation (6.5.9) suggests that (6.5.8) is the correct form for the Lorentz force in four-vector notation, one might nevertheless query whether the right-hand side of equation (6.5.8) is the most general form that one should consider if the requirement is merely that both sides of the equation should behave like a four-vector. Consider the relation

$$F_\alpha = m_0 \frac{dU_\alpha}{d\tau} = Q \sum_\beta F_{\alpha\beta} U_\beta + C U_\alpha \tag{6.5.10}$$

where C is some function which has yet to be determined. Now both sides of the equation are four-vectors and since $U = (0, ic)$ in the limit $\mathbf{u} \to 0$ it is evident that the above equation still satisfies (6.5.1) in the limit $\mathbf{u} \to 0$. However, if we multiply both sides by U_α and sum over α we find

$$m_0 \sum_\alpha U_\alpha \frac{dU_\alpha}{d\tau} = Q \sum_{\alpha\beta} U_\alpha F_{\alpha\beta} U_\beta + C \sum_\alpha U_\alpha^2 = -c^2 C$$

where we have used the fact that

$$\sum_{\alpha\beta} U_\alpha F_{\alpha\beta} U_\beta = 0$$

since $F_{\alpha\beta}$ is antisymmetric (6.2.35), and that $\sum_\alpha U_\alpha^2 = -c^2$ by (4.2.5). But equation (4.2.5) also implies that

$$2 \sum_\alpha U_\alpha \frac{dU_\alpha}{d\tau} = \frac{d}{d\tau} \sum_\alpha U_\alpha^2 = -\frac{dc^2}{d\tau} = 0$$

Hence we conclude that $C = 0$ in (6.5.10), and so our original formulation in (6.5.8) was correct. However, the above argument shows that in formulating the correct covariant equations we must demand that *not only does the equation satisfy the correct Lorentz transformation properties but that it must also satisfy the appropriate physical constraints*. In the present situation we introduced two constraints

(1) that the equation for the Lorentz force reduced to the electrostatic equation (6.5.1) in the limit $\mathbf{u} \to 0$,

(2) the fact that $\sum_\alpha U_\alpha^2 = -c^2$ and was therefore a constant.

6.5.2   The magnetic deflection of charged particles

The equation for the Lorentz force appears repeatedly in atomic and nuclear physics. Historically its most famous application was in the experiments of J. J. Thomson on the determination of e/m for electrons and positive ions; nowadays the same principle is employed in the design of television tubes. This work, however, is in the limit $\gamma_u \to 1$. In the large particle accelerators and their ancillary apparatus $\gamma_u \gg 1$. Let us examine the behaviour of a particle of charge $Q = e$ in a large accelerator. Here a magnetic field $\mathbf{B}$ alone is applied to constrain the particles and they are injected at right angles to $\mathbf{B}$; since $\mathbf{E} = 0$ equation (6.5.9) becomes

$$\mathbf{f} = e\mathbf{u} \times \mathbf{B} = euB\mathbf{n} \qquad (6.5.11)$$

where

$$u = |\mathbf{u}| \qquad B = |\mathbf{B}|$$

and we have introduced a unit vector $\mathbf{n}$ perpendicular to both $\mathbf{u}$ and $\mathbf{B}$ (figure 6.2). Since $\mathbf{f}$ is perpendicular to $\mathbf{u}$, it follows that

$$\mathbf{u} \cdot \mathbf{f} = 0 \qquad (6.5.12)$$

The force experienced by the particle can be expressed in terms of acceleration in the manner of (4.9.2)

$$\mathbf{f} = m \frac{d\mathbf{u}}{dt} + \mathbf{u} \frac{dm}{dt}$$
$$= \gamma_u m_0 \frac{d\mathbf{u}}{dt} + \frac{\mathbf{u}}{c^2} \frac{dE}{dt} \qquad (6.5.13)$$

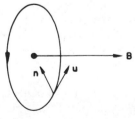

Figure 6.2   Motion of a charged particle in a magnetic field. The plane containing the velocity $\mathbf{u}$ is perpendicular to $\mathbf{B}$, and $\mathbf{n}$ is a unit vector lying in this plane and perpendicular to $\mathbf{u}$

where m_0 is the rest mass of the particle and the second line follows from the Einstein equations ((4.2.8) and (4.2.11)). Now we have shown in (6.5.2) that

$$\frac{dE}{dt} = \mathbf{u} \cdot \mathbf{f}$$

and since $\mathbf{u} \cdot \mathbf{f} = 0$ in the present circumstances (6.5.12), the Lorentz force equation (6.5.11) becomes

$$\gamma_u m_0 \frac{d\mathbf{u}}{dt} = euB\mathbf{n}$$

Now the relations (6.5.12) and (6.5.2)

$$\mathbf{u} \cdot \mathbf{f} = \frac{dE}{dt} = m_0 c^2 \frac{d\gamma}{dt} = m_0 c^2 \frac{d}{dt} \frac{1}{\sqrt{(1 - u^2/c^2)}} = 0$$

implies that $\mathbf{u}$ must be constant in magnitude and that its direction only is changing as a function of time. Since $\mathbf{n}$ is always perpendicular to $\mathbf{u}$, the particle is therefore rotating with an angular velocity ω_c given by

$$\omega_c = \frac{d\theta}{dt} = \frac{du}{udt} = \frac{e}{\gamma_u m_0} B\mathbf{n}$$

or

$$\omega_c = \frac{e}{\gamma_u m_0} B \qquad (6.5.14)$$

ω_c is known as the cyclotron frequency. It is apparent from this equation that ω_c falls steadily from the classical limit eB/m_0 as the velocity increases. This important factor must be allowed for in the design of the frequency of the accelerating voltage in the large proton synchrotrons.

Equation (6.4.15) also determines the radius R of the orbit of the particle: Since $R\omega_c = u$ we find that*

$$R = \frac{u}{\omega_c} = \frac{m_0 \gamma_u u}{eB} = \frac{p}{eB} \qquad (6.5.15)$$

where $p = |\mathbf{p}|$ is the linear momentum of the particle. For economic reasons it is desirable to keep R fixed in the design of large accelerators; hence as the energy, and therefore momentum, of the particle increases, so also must the magnetic field B. Equation (6.5.15) also is used in the kinematic analysis of bubble chamber pictures of high-energy particle reactions—the measurements of the radius of curvature of tracks of elementary particles in pictures like figure 4.16 allow the momentum of these particles to be determined.

---

* It is amusing to notice that starting from the equation of electrostatic force in (6.5.1), the principle of the Lorentz covariance of physical equations has led us to the radius of curvature of particles in a magnetic field.

6.6 The precession of the polarisation of particles moving in a homogeneous electromagnetic field [*]

6.6.1 Construction of the equation of motion

In this section we shall discuss the theory which lies behind an important series of experiments to test the validity of quantum electrodynamics. The theory is concerned with the precession of the 'spin' of a particle in a homogeneous magnetic field. We shall describe its importance later.

In classical nonrelativistic electrodynamics a magnetic dipole of strength μ in a magnetic field of strength $\mathbf{B}$ experiences a torque [†]

$$\mathbf{N} = \mu \times \mathbf{B}$$

which for an elementary particle can be interpreted as

$$\mathbf{N} = \frac{d\mathbf{S}}{dt} = \mu \times \mathbf{B} = \frac{ge}{2m_0} \mathbf{S} \times \mathbf{B} \qquad (6.6.1)$$

where $\mathbf{S}$ is the spin of the particle in its rest frame, μ its magnetic moment, e its charge, m_0 its rest mass and g the Landé factor (the gyromagnetic ratio, which is, in fact, defined by this equation).

Our purpose is to turn the above equation into a relativistically invariant expression. Since g, e and m_0 are constants, we concentrate on $\mathbf{S}$, t and $\mathbf{B}$. Now we have seen in chapter 5 we can construct a four-vector Γ (the Pauli–Lubanski vector) which in the nonrelativistic limit (5.3.5) gives

$$\Gamma_R = (m_0 c \mathbf{S}, 0)$$

In that chapter we did not define a 'spin' as $\Gamma / m_0 c$ because of the problems which then arise for $m_0 = 0$. Here we are examining particles of finite rest mass m_0, and so it is safe to introduce a 'spin' four-vector

$$\hat{S} = \frac{\Gamma}{m_0 c} \qquad (6.6.2)$$

which in the nonrelativistic limit (that is, $u/c \to 0$) gives

$$\hat{S}_R = (\mathbf{S}, 0) = (\hat{S}_R, 0) \qquad (6.6.3)$$

where the second form is introduced in order to avoid a confusing notation later. The scalar product of $\hat{S}$ with the momentum four-vector still satisfies the relation (5.3.7)

$$\sum_{\sigma=1}^{4} \hat{S}_\sigma p_\sigma = 0$$

[*] BARGMANN, V., MICHEL, L., and TELEGDI, V. L., *Phys. Rev. Lett.* **2**, 435, 1959.
[†] See for example, §7–11 of *Classical Electricity and Magnetism* 2nd edition, by PANOFSKY, W. K. H., and PHILLIPS, M., Addison–Wesley, 1964.

and since $p_\sigma = m_0 U_\sigma$ (4.2.7), where U is the four-velocity (4.2.3), we can also write

$$\sum_\sigma \hat{S}_\sigma U_\sigma = \gamma_u (\hat{\mathbf{S}} \cdot \mathbf{u} + ic\hat{S}_4) = 0 \qquad (6.6.4)$$

Now consider dS/dt in (6.6.1). In forming a suitable relativistic notation we may proceed in analogy with the four-acceleration in the Lorentz force equation (§6.5.1). There the acceleration du/dt of equation (6.5.1) became $dU_\alpha/d\tau$ in (6.5.8). In a similar manner we make the substitution

$$\frac{dS}{dt} \rightarrow \frac{d\hat{S}_\alpha}{d\tau}$$

Next consider the right-hand side of (6.6.1). We require a combination of four-vectors and tensors with the index α unsaturated, which in the limit $u/c \rightarrow 0$ reduce to equation (6.6.1). An inspection of this equation together with our experience of the construction of the Lorentz force equation (§6.5.1), suggests that the required four-vectors and tensors will be $\hat{S}$, F and U. The requirements of dimensional analysis imply that the equations be linear in S and F; no such restriction exists for U since we can always add c (in fact the four-velocity U is sometimes defined so that $\sum_\sigma U_\sigma^2 = -1$). Possible combinations are

$$\sum_\beta F_{\alpha\beta} S_\beta \qquad U_\alpha \sum_{\beta\gamma} U_\beta F_{\beta\gamma} \hat{S}_\gamma \qquad (6.6.5)$$

All other combinations either vanish because of the relation $\sum_\sigma \hat{S}_\sigma U_\sigma = 0$ or the antisymmetry of $F_{\alpha\beta}$.

Thus the equation (6.6.1) becomes, in four-vector notation

$$\frac{d\hat{S}_\alpha}{d\tau} = A \sum_\beta F_{\alpha\beta} \hat{S}_\beta + B U_\alpha \sum_{\beta\gamma} U_\beta F_{\beta\gamma} \hat{S}_\gamma \qquad (6.6.6)$$

where A and B are constants to be determined.

The value for A may be ascertained by reducing (6.6.6) to the rest frame; in this frame U (4.2.6) becomes

$$U_\alpha = 0, 0, 0, ic$$

and so the space parts of $\hat{S}_\alpha$ are

$$\frac{d\hat{S}_{jR}}{dt} = A \sum_k F_{jk} \hat{S}_{kR} \qquad j = 1, 2, 3$$

If we consider $j = 1$ we find from (6.2.36)

$$\begin{aligned}
\frac{d\hat{S}_{1R}}{dt} &= A(F_{12}\hat{S}_{2R} + F_{13}\hat{S}_{3R}) \\
&= A(B_3\hat{S}_{2R} - B_2\hat{S}_{3R}) \\
&= A(\hat{\mathbf{S}}_R \times \mathbf{B})_1
\end{aligned}$$

hence a comparison with (6.6.1) and (6.6.3) gives

$$A = \frac{ge}{2m_0}$$

The value of B in (6.6.6) may be established by first noting that equation (6.6.4) implies

$$\sum_\alpha U_\alpha \frac{d\hat{S}_\alpha}{d\tau} = -\sum_\alpha \hat{S}_\alpha \frac{dU_\alpha}{d\tau} = -\frac{e}{m_0} \sum_{\alpha\beta} \hat{S}_\alpha F_{\alpha\beta} U_\beta$$

$$= \frac{e}{m_0} \sum_{\alpha\beta} U_\alpha F_{\alpha\beta} \hat{S}_\beta$$

where we have used the Lorentz force equation (6.5.8) with e substituted for Q; the last equation follows from the antisymmetry of $F_{\alpha\beta}$ (6.2.35). If we therefore multiply (6.6.6) by U_α and sum over α we find

$$\sum_\alpha U_\alpha \frac{d\hat{S}_\alpha}{d\tau} = \frac{ge}{2m_0} \sum_{\alpha\beta} U_\alpha F_{\alpha\beta} \hat{S}_\beta - Bc^2 \sum_{\beta\gamma} U_\beta F_{\beta\gamma} \hat{S}_\gamma$$

$$= \frac{e}{m_0} \sum_{\alpha\beta} U_\alpha F_{\alpha\beta} \hat{S}_\beta$$

where we have used the relation $\sum_\alpha U_\alpha^2 = -c^2$ (4.2.5).

Now despite the difference in subscripts all the summations cover 1 to 4 in the above equation and so

$$Bc^2 = \frac{ge}{2m_0} - \frac{e}{m_0}$$

or

$$B = \frac{e}{2m_0 c^2}(g - 2)$$

Thus equation (6.6.6) becomes

$$\frac{d\hat{S}_\alpha}{d\tau} = \frac{e}{2m_0 c^2}\left[gc^2 \sum_\beta F_{\alpha\beta} \hat{S}_\beta + (g - 2) U_\alpha \sum_{\beta\gamma} U_\beta F_{\beta\gamma} \hat{S}_\gamma \right] \tag{6.6.7}$$

One interesting consequence immediately follows from this equation; if $g = 2$ we find

$$\frac{d\hat{S}_\alpha}{d\tau} = \frac{e}{m_0} \sum F_{\alpha\beta} \hat{S}_\beta$$

which is the same equation of motion as for the four-velocity (6.5.8) with S substituted for U and e for Q. Thus for a particle rotating in a plane perpendicular to a magnetic field (that is, the physical conditions of (§6.5.2) we might intuitively guess that the 'spin' rotates with the same frequency as the particle. In the next section we shall show this conclusion to be correct.

6.6.2 Solutions for the equation of motion *
Equation (6.6.7)

$$\frac{d\hat{S}_\alpha}{d\tau} = \frac{e}{2m_0c^2} \left[gc^2 \sum_\beta F_{\alpha\beta}\hat{S}_\beta + (g-2) \, U_\alpha \sum_{\beta\gamma} U_\beta F_{\beta\gamma}\hat{S}_\gamma \right]$$

represents the most general form in which we can express the equation of motion of the 'spin' in a homogeneous electromagnetic field. In this section we shall first show that the magnitude of $\hat{S}$ remains invariant; that is, the degree of polarisation stays unchanged as a function of time. Only the direction of $\hat{S}$ can change.

The invariance of the magnitude is easily demonstrated. If we multiply both sides of (6.6.7) by $\hat{S}_\alpha$ and sum over α we find for the left-hand side

$$\sum_\alpha S_\alpha \frac{dS_\alpha}{d\tau} = \frac{1}{2} \frac{d}{d\tau} \left(\sum_\alpha \hat{S}_\alpha^2 \right) = \frac{1}{2} \frac{d}{d\tau} (\hat{S}_R^2)$$

where the last equation follows from the Lorentz invariance of $\sum_\alpha \hat{S}_\alpha^2$ (6.6.3). Now consider the right-hand side; we find immediately that

$$\sum_{\alpha\beta} \hat{S}_\alpha F_{\alpha\beta}\hat{S}_\beta = 0$$

$$\sum_\alpha \hat{S}_\alpha U_\alpha = 0$$

where the first result follows from the antisymmetric nature of F; that is, $F_{\alpha\beta} = -F_{\beta\alpha}$ (6.2.35), and the second result was given in (6.6.4). Thus we have shown that

$$\frac{d}{d\tau} (\hat{S}_R^2) = 0 \tag{6.6.8}$$

that is the 'length' of the 'spin' vector remains unchanged with time. Thus the degree of polarisation of a beam of particles remains unaltered when it traverses an electromagnetic field. As we shall see below, the direction changes. Our next task is to determine this change.

Since $\hat{S}$ is defined as Γ/m_0c (6.6.2), it is not difficult to see that we may adapt equations (5.4.1) and (5.4.4) to relate $\hat{S}$ in any reference frame to the particle's rest frame

$$\hat{S} = s(\gamma \mathbf{l} \cos\theta_R + \mathbf{n} \sin\theta_R)$$
$$\hat{S}_4 = i\gamma\beta s \cos\theta_R \tag{6.6.9}$$

here $s^2 = S^2 = \hat{S}_R^2$ (§ 5.4 and (6.6.3)), and as before $\mathbf{l}$ and $\mathbf{n}$ are unit orthogonal

---

* In this section we follow the discussion of Hagedorn in *Relativistic Kinematics*, Benjamin, 1963.

vectors defining the plane in which $\hat{\mathbf{S}}_R$ lies

$$\mathbf{l}=\mathbf{u}/u \qquad \mathbf{l}\cdot\mathbf{n}=0 \qquad \mathbf{l}^2=\mathbf{n}^2=1$$

Since the direction of $\mathbf{u}$ changes as a function of time, and $\mathbf{l}$ and $\mathbf{n}$ define the direction of $\hat{\mathbf{S}}_R$ with respect to $\mathbf{u}$ at any instant in the particle's rest frame, the directions of $\mathbf{l}$ and $\mathbf{n}$ continually change in a reference frame in which the particle is moving. The unit vectors $\mathbf{l}$ and $\mathbf{n}$ can be regarded as the nonrelativistic limits of two four-vectors L and N which possess components

$$L=(\gamma_u \mathbf{l}, \, i\gamma_u \beta) \qquad N=(\mathbf{n},\, 0) \qquad (6.6.10)$$

in the reference frame in which the particle is moving. The forms for L and N are readily constructed with the Lorentz boost $L(\beta)$ as used for Γ in equations (5.4.1).

The four-vectors L and N allow us to reformulate equations (6.6.9) as

$$\hat{\mathbf{S}}=s(\mathbf{L}\cos\theta_R+\mathbf{N}\sin\theta_R)$$
$$\hat{S}_4=s(L_4\cos\theta_R+0)$$

that is, the space and time components of a four-vector

$$\hat{S}=s(L\cos\theta_R+N\sin\theta_R) \qquad (6.6.11)$$

The definitions of L and N in (6.6.10) also lead to the following equations

$$\sum_\alpha L_\alpha^2=\sum_\alpha N_\alpha^2=1$$
$$\sum_\alpha N_\alpha L_\alpha=\sum_\alpha L_\alpha U_\alpha=\sum_\alpha N_\alpha U_\alpha=0 \qquad (6.6.12)$$
$$\sum_\alpha \frac{dL_\alpha}{d\tau} L_\alpha=\sum_\alpha \frac{dN_\alpha}{d\tau} N_\alpha=0$$

Equations (6.6.11) and (6.6.12) together lead to a solution of the equation of motion (6.6.7). If we multiply both sides of (6.6.7) by N_α and sum over α, the left-hand side gives

$$\sum_\alpha N_\alpha \frac{d\hat{S}_\alpha}{d\tau}=s\sum_\alpha\left[N_\alpha \frac{dL_\alpha}{d\tau}\cos\theta_R+N_\alpha \frac{dN_\alpha}{d\tau}\sin\theta_R\right.$$
$$\left.+\frac{d\theta_R}{d\tau}(N_\alpha N_\alpha \cos\theta_R-N_\alpha L_\alpha \sin\theta_R)\right]$$
$$=s\sum_\alpha\left(N_\alpha \frac{dL_\alpha}{d\tau}+\frac{d\theta_R}{d\tau}\right)\cos\theta_R \qquad (6.6.13)$$

When we give the same treatment to the right-hand side of (6.6.7) the second term vanishes since $\sum_\alpha N_\alpha U_\alpha=0$, whilst $\sum_{\alpha\beta} N_\alpha F_{\alpha\beta} N_\beta$ disappears in the first term because of the antisymmetric properties of $F_{\alpha\beta}$ (6.2.35). We are then left with

$$\sum_\alpha N_\alpha \frac{d\hat{S}_\alpha}{d\tau} = \frac{e}{2m_0 c^2} gc^2 s \sum_{\alpha\beta} N_\alpha F_{\alpha\beta} L_\beta \cos\theta_R$$

and upon equating this equation with (6.6.13) we obtain

$$\frac{d\theta_R}{d\tau} = \frac{ge}{2m_0} \sum_{\alpha\beta} N_\alpha F_{\alpha\beta} L_\beta - \sum_\alpha N_\alpha \frac{dL_\alpha}{d\tau} \qquad (6.6.14)$$

Our next task is to examine the final term. If we recall that $N=(\mathbf{n}, 0)$ and that $\mathbf{l}\cdot\mathbf{n}=0$ then we can write

$$\sum_\alpha N_\alpha \frac{dL_\alpha}{d\tau} = \mathbf{n}\cdot\frac{d\mathbf{L}}{d\tau} = \mathbf{n}\cdot\frac{d}{d\tau}(\gamma_u \mathbf{l}) = \gamma_u \mathbf{n}\cdot\frac{d\mathbf{l}}{d\tau}$$

but a similar equation also exists for the four-velocity U

$$\sum_\alpha N_\alpha \frac{dU_\alpha}{d\tau} = \mathbf{n}\cdot\frac{d\mathbf{U}}{d\tau} = \mathbf{n}\cdot\frac{d}{d\tau}(\gamma_u \mathbf{l}u) = \gamma_u u \mathbf{n}\cdot\frac{d\mathbf{l}}{d\tau}$$

hence, with the aid of the equation for the Lorentz force (6.5.8), we find that

$$\sum_\alpha N_\alpha \frac{dL_\alpha}{d\tau} = \frac{1}{u} \sum_\alpha N_\alpha \frac{dU_\alpha}{d\tau}$$

$$= \frac{e}{m_0 u} \sum_{\alpha\beta} N_\alpha F_{\alpha\beta} U_\beta$$

and so (6.6.14) becomes

$$\frac{d\theta_R}{d\tau} = \frac{ge}{2m_0} \sum_{\alpha\beta} N_\alpha F_{\alpha\beta} L_\beta - \frac{e}{m_0 u} \sum_{\alpha\beta} N_\alpha F_{\alpha\beta} U_\beta \qquad (6.6.15)$$

The final stage in the analysis is to replace $F_{\alpha\beta}$ by the measurable fields $\mathbf{E}$ and $\mathbf{B}$ and to free the equation from the X, Y, Z coordinate frame. Since $N_4 = 0$ we can write

$$\sum_\alpha N_\alpha F_{\alpha\beta} = \sum_i n_i F_{i\beta} \qquad i=1, 2, 3$$

and, as examples, for $\beta=1, 4$ we find from (6.2.36)

$$\beta=1 \quad \sum_\alpha N_\alpha F_{\alpha\beta} = n_1 F_{11} + n_2 F_{21} + n_3 F_{31}$$

$$= -n_2 B_3 + n_3 B_2$$

$$= (\mathbf{B}\times\mathbf{n})_1$$

$$\beta = 4 \qquad \sum_{\alpha} N_{\alpha} F_{\alpha\beta} = n_1 F_{14} + n_2 F_{24} + n_3 F_{34}$$

$$= -\frac{i}{c} (n_1 E_1 + n_2 E_2 + n_3 E_3)$$

$$= -\frac{i}{c} \mathbf{n} \cdot \mathbf{E}$$

The four-vectors L and U in (6.6.15) can be written as

$$L = (\gamma_u \mathbf{l}, \, i\gamma_u \beta); \qquad U = (\gamma_u \mathbf{u}, \, ic\gamma_u)$$

and so equation (6.6.15) finally becomes

$$\frac{d\theta_R}{d\tau} = \gamma_u \frac{d\theta_R}{dt} = \frac{ge}{2m_0} \left[(\mathbf{B} \times \mathbf{n}) \cdot \mathbf{l}\gamma_u + \mathbf{n} \cdot \mathbf{E}\gamma_u \beta/c \right]$$

$$\frac{-e}{m_0 u} \left[(\mathbf{B} \times \mathbf{n}) \cdot \mathbf{u}\gamma_u + \mathbf{n} \cdot \mathbf{E}\gamma_u \right]$$

or

$$\frac{d\theta_R}{dt} = \frac{e}{2m_0} \left[(g-2)(\mathbf{B} \times \mathbf{n}) \cdot \mathbf{l} + \left(g\frac{\beta}{c} - \frac{2}{u} \right) \mathbf{n} \cdot \mathbf{E} \right] \qquad (6.6.16)$$

This is the equation of Bargmann, Michel and Telegdi for the precession of 'spin' in an electromagnetic field. The reader will have noticed that the 'length' of the spin vector disappeared at the stage of equation (6.6.14), and we have been left with an equation for the rate of change of direction of the 'spin', that is, for the polarisation vector. The equation is valid for arbitrary spins (in the quantum mechanical sense of $S = \frac{1}{2}\hbar, \hbar, \ldots$) since it connects with quantum mechanics via Ehrenfest's theorem. It has been checked by quantum mechanical treatments for $S = \frac{1}{2}\hbar$.

6.6.3   The 'g − 2' experiment

A situation of particular interest is that where $\mathbf{E}$ is zero and $\mathbf{B}$ is perpendicular to the velocity $\mathbf{u}$

$$\mathbf{E} = 0 \qquad (\mathbf{B} \times \mathbf{n}) \cdot \mathbf{l} = B \qquad (6.6.17)$$

We have already examined the motion of the particle for these conditions in §5.2; it was found that the particle moved in a circle whose plane was perpendicular to the direction of $\mathbf{B}$ and that the frequency of rotation was (compare (6.5.14))

$$\omega_c = \frac{e}{m_0 \gamma_u} B$$

The conditions (6.6.17) imply that equation (6.6.16) becomes

$$\frac{d\theta_R}{dt} = \omega_a = \frac{e}{2m_0} (g-2) B = \gamma_u \left(\frac{g-2}{2} \right) \omega_c \qquad (6.6.18)$$

(The subscript a refers to the so-called anomalous angular velocity which exists for $g \neq 2$.)Let us recall that θ_R defines the direction of polarisation of the particle *relative to the velocity vector* in the particle's rest frame. Thus for $g=2$, $\omega_a = 0$, and the polarisation vector follows the velocity vector, as we anticipated in §6.6.1. This behaviour is displayed in figure 6.3(a). For $g \neq 2$ the polarisation vector rotates relative to the velocity vector with an 'anomalous' angular velocity ω_a which *is independent of* γ_u. Thus the precession period T_a of the anomalous moment defined by

$$\omega_a T_a = 2\pi$$

is the same as in the nonrelativistic limit.

If we define the cyclotron period by

$$\omega_c T_c = 2\pi$$

we find, however, that the angular change per rotation of the particle is dependent on γ_u, since

$$\Delta\theta_R = \omega_a T_c = \omega_a \frac{2\pi}{\omega_c} = 2\pi\gamma_u \left(\frac{g-2}{2}\right) \tag{6.6.19}$$

Thus the change in angle per rotation increases with γ_u. This behaviour of the polarisation vector is illustrated in figure 6.3(b). We assume $g > 2$ and that

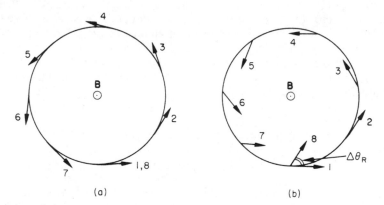

(a)   (b)

Figure 6.3 Behaviour of the polarisation direction for $g = 2$ (a) and $g > 2$ (b). The arrows indicate the polarisation vector in the time sequence 1 to 8

the polarisation is initially parallel to the velocity vector at the position 1. It is apparent from the diagram that after several rotations of the particle θ_R will become 2π so that it is again parallel to **u**, and later the angle will open up again. Thus as a function of time the polarisation vector will oscillate about **u** with an angular velocity ω_a.

The principle of the oscillation of the polarisation vector about the velocity vector in a uniform magnetic field for $g \neq 2$ has been exploited in a series of increasingly accurate experiments performed at the CERN laboratory in Geneva during the past 10 years. The subject of quantum electrodynamics enables the g value for the μ-particles to be calculated to an extraordinary degree of accuracy, and accordingly the measurement of $g-2$ to a comparable degree of accuracy is regarded as one of the most sensitive tests of the validity of quantum electrodynamics.

The principle of the experiment is illustrated in figure 6.4*. A beam of

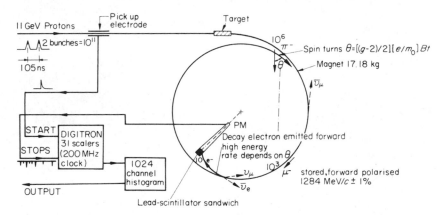

Figure 6.4 Schematic view of $g-2$ experiment. The arrival of the proton bunches are electronically recorded from the pick-up electrode and START the timing device. The protons produce pions at the target, these decay to muons and in turn to electrons — the latter are detected by the lead scintillator sandwich and photo-multiplier (P.M.) and a STOP signal is sent to the electronic timer.

protons from the CERN proton synchrotron strikes a target which produces π mesons. The pions decay yielding muons and neutrinos (v)

$$\pi \rightarrow \mu v$$

and some of these muons follow the direction of the pions and are captured into a circular orbit in a magnet of field strength 17.18 kilogauss. The muons move in an evacuated tube and so could remain rotating indefinitely. However as time passes they decay into electrons and neutrinos

$$\mu \rightarrow e v \bar{v}$$

Now both the pion and muon decays are weak interactions, and they therefore violate the law of parity conservation, which we discussed in §5.6.1. As a consequence of this violation, the spins of the muons are polarised along their direction of motion, and the decay electrons emerge preferentially along the direction of the muon spin. The decay electrons are detected electronically (figure 6.4).

---

* A very readable account of the experiment may be found in an article by J. Bailey and E. Picasso, *Progress in Nuclear Physics*, **12**, 43, 1970.

The time intervals between the arrival of the protons and the appearance of the decay electrons are electronically measured. These times occur in a distribution of the form

$$N(t) = C[1 - A\cos(\omega_a t + \phi)]\exp(-t/\gamma_\mu \tau_0)$$

where C, A and ϕ are constants peculiar to the apparatus and τ_0 is the mean lifetime of the muon in its rest frame. In principle, one should also include a

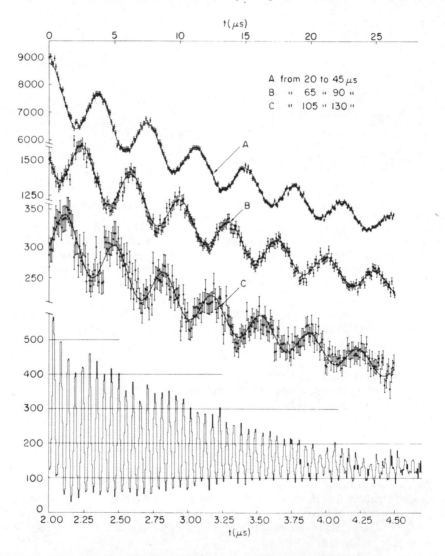

Figure 6.5 Distribution of decay electrons as a function of time (upper curves). The bottom curve shows the rotation frequency of the muon at early time.

term for the mean lifetime of the pion—this is $\sim 1/100$ of τ_0 and essentially affects the first μ s of $N(t)$. It can therefore be ignored. The results are displayed in figure 6.5. An analysis of the curves yielded values for $\gamma\tau_0$ and ω_a. The first quantity is important for checking relativistic time dilatation and the twin paradox (§§ 3.1 and 3.5), whilst ω_a yields $(g-2)/2$ via equation (6.6.18). The experimental result is

$$\frac{(g-2)_{\text{exp}}}{2} = (116616 \pm 31) \times 10^{-8}$$

whilst the theoretical value is $(116588.0) \times 10^{-8}$. The difference

$$\frac{(g-2)_{\text{exp}}}{2} - \frac{(g-2)_{\text{th}}}{2} = (28 \pm 31) \times 10^{-8} \equiv (240 \pm 270) \text{ ppm}$$

appears to be negligible, thus providing a valuable check on the validity of quantum electrodynamics. A further experiment, which utilises both electric and magnetic fields is being planned, however. With this new experiment it is hoped to improve the accuracy of the determination of $(g-2)$ result by at least an order of magnitude.

PROBLEMS

6.1 Explain the physical significance of the negative sign in equation (6.1.13).

6.2 Check that the determinant of L and L^{-1} is one for equations (6.2.18) and (6.2.22) respectively.

6.3 Check equations (6.2.33), and show that Maxwell's equations are invariant in $\mathbf{E}$ and $\mathbf{B}$ notation.

6.4 Consider an electromagnetic field that is pure magnetic in a reference frame Σ. Show that in a reference frame Σ' moving past with velocity $\mathbf{v}$ an observer records an electric field given by

$$\mathbf{E}' = (\mathbf{v} \times \mathbf{B})_\perp$$

6.5 Construct equation (6.2.36), and hence show that Maxwell's equation can also be written as

$$\sum_\beta \frac{\partial F_{\alpha\beta}}{\partial x_\beta} = j_\alpha$$

$$\frac{\partial F_{\alpha\beta}}{\partial x_\gamma} + \frac{\partial F_{\beta\gamma}}{\partial x_\alpha} + \frac{\partial F_{\gamma\alpha}}{\partial x_\beta} = 0$$

6.6 Show that

$$\sum_\lambda \frac{\partial^2}{\partial x_\lambda^2} \neq \sum_\lambda \frac{\partial^2}{\partial x_\lambda'^2}$$

for a Galilean transformation (1.1.2).

6.7 Show that the exponent in equation (6.3.12) is Lorentz invariant.

6.8 Show that $\mathbf{E}^2 - c^2\mathbf{B}^2$ and $\mathbf{E}\cdot\mathbf{B}$ are Lorentz invariants, and that for free electromagnetic waves ($j_\alpha = 0$) $\mathbf{E}$ is perpendicular to $\mathbf{B}$ in every reference frame.

6.9 Show that the Lorentz force equation (6.5.9) can be obtained from the electrostatic equation (6.5.1) by a suitable Lorentz transformation.

6.10 The invariant amplitude, T, for the decay of an ω meson to three π mesons can be represented by

$$T = \sum_{\alpha\beta\gamma\delta} \varepsilon_{\alpha\beta\gamma\delta} \hat{S}_\alpha k_{A\beta} k_{B\gamma} k_{C\delta}$$

where ε is the antisymmetric tensor of equation (5.2.14), the labels A, B and C refer to the three π mesons and $\hat{S}_\alpha$ is the spin polarisation vector of the ω meson. Show that in the rest frame of the ω particle the above relation reduces to

$$T = 3\frac{i}{c} M_\omega \hat{S}_R \cdot k_A \times k_B$$

where M_ω is the mass of the ω meson. If $|T|^2$ represents the probability for ω decay show that this function reduces to zero when B and C are both antiparallel to A.

6.11 Show that $\sum_{\alpha\beta} U_\alpha F_{\alpha\beta} U_\beta = 0$.

6.12 How many times must a beam of muons of momentum 1.284 GeV/c rotate in a magnetic field to produce transverse polarisation from an initial state of longitudinal polarisation? (Mass of muon $= 106$ MeV/c^2; $(g-2)/2$ for muon $= 1.2 \times 10^{-3}$).

6.13 What solutions would be found in §6.5.2 and 6.6.3 if $E=0$ and $\mathbf{B}$ and $\mathbf{u}$ are parallel?

Appendixes

A.1 The Michelson–Morley experiment

The essential features of the apparatus first devised by Michelson[*], and then improved upon in collaboration with Morley[†] are displayed in figure A.1.1.

Light from a monochromatic source S falls on a glass plate A, which has a semitransparent metal coating on its front face. The coating causes the light from S to split into two beams which travel perpendicularly to each other to mirrors B and C, where they are reflected. Some of the reflected light from both beams then travels to the telescope T. A compensating plate P between A and C, ensures that the light paths A→C→A→T and A→B→A→T traverse the same thickness of glass in the interests of optical symmetry.

The mirrors A and B are slightly inclined away from the 90° position with respect to each other. The light source S produces monochromatic light and

---

[*] MICHELSON, A. A., *Am. J. Sci.* **122**, 120, 1881.
[†] MICHELSON, A. A., and MORLEY, E. W., *Am. J. Sci.* **134**, 333, 1887.

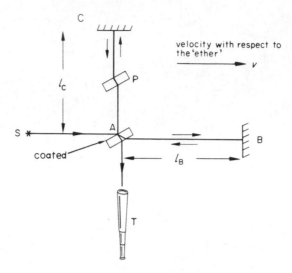

Figure A.1.1

so straight line interference fringes are observed in T, since the system is equivalent to a thin optical wedge ('the Michelson interferometer')' Thus if the optical path lengths $A \to B \to A$, $A \to C \to A$ are $2l_B$ and $2l_C$ respectively the condition for constructive interference is

$$2(l_B - l_C) = n\lambda \qquad (A.1.1)$$

where n is an integer.

Now let us assume that the apparatus is moving in the direction AB with respect to the 'ether reference frame'. As we discussed in §1.3 the time for the light to travel back and forward along l_B and l_C is respectively

$$t_B = \frac{l_B}{c+v} + \frac{l_B}{c-v} = \frac{2l_B}{c(1-v^2/c^2)}$$

$$t_C = \frac{2l_C}{\sqrt{(c^2-v^2)}} = \frac{2l_C}{c\sqrt{(1-v^2/c^2)}}$$

The time difference, δt, between the two periods, if $v \ll c$, is then

$$\delta t = t_B - t_C \sim \frac{2l_B}{c}(1 + v^2/c^2) - \frac{2l_C}{c}(1 + v^2/2c^2)$$

$$= \frac{2(l_B - l_C)}{c} + \frac{v^2}{c^3}(2l_B - l_C) \qquad (A.1.2)$$

Now let the apparatus be swung through 90° so that AC points along the direction of motion. The time difference now becomes

$$\delta t' = t'_B - t'_C \sim \frac{2l_B}{c}(1 + v^2/2c^2) - \frac{2l_C}{c}(1 + v^2/c^2)$$

$$= \frac{2(l_B - l_C)}{c} + \frac{v^2}{c^3}(l_B - 2l_C) \tag{A.1.3}$$

Thus $\delta t \neq \delta t'$, and the change in time difference should cause a shift in the interference pattern by n fringes, where

$$\delta n = \frac{c(\delta t - \delta t')}{\lambda} = \frac{(l_B + l_C)v^2}{c^3} \tag{A.1.4}$$

In Michelson's first experiment $l_B = l_C \sim 1.2$ m and $\lambda \sim 6 \times 10^{-7}$ m. A suitable first approximation for v is the velocity of the earth in its orbit; this is roughly 30 km/s and so $v/c \sim 10^{-4}$ giving $n = 0.04$ fringe. Although this displacement of the fringe pattern is small, it was nevertheless detectable. Michelson set an upper limit of $n = 0.02$ from his experiment, and concluded that 'The result of the hypothesis of a stationary ether is thus shown to be incorrect'.

The improved version of the experiment by Michelson and Morley determined an upper limit of 0.01 fringes where 0.4 might have been expected. Another way of stating this result is to say that the experiment yields an upper limit v_u on the velocity of the earth through the ether, since if we denote the observed upper limit on the fringe shift by δn_u and that calculated for $v = 30$ km/s by δn_{30}, then from (A.1.4)

$$v_u = 30 \sqrt{\left(\frac{\delta n_u}{\delta n_{30}}\right)} \text{ km/s}$$

The figures of 0.4 and 0.01 for δn_{30} and δn_u respectively thus lead to the result $v_u \leqslant 5$ km s^{-1}.

Many experiments have been performed since 1887, and the net result has been to make v_u progressively smaller. The subsequent optical experiments have been listed by Shankland et al*; they have led to the results $v_u \leqslant 1.5$ km/s. Moreover, in recent years experiments have been performed using the techniques associated with masers[†] and the Mössbauer effect[‡]. The net results from these experiments are given in table A.1.

Table A.1

Technique	v_u
Michelson–Morley (1887)	5 km/s
ditto by Joos (1930)	1.5 km/s
Masers	3×10^{-2} km/s
Mössbauer effect	5×10^{-5} km/s

* SHANKLAND, R. S., McCUSKEY, S. W., LEONE, F. C., and KNERTI, G., Rev. mod. phys. 27, 167, 1955.
† CEDARHOLM, J. P., BLAND, G. F., HAVENS, B. L., and TOWNES, C. H., Phys. Rev. Letters 1, 342, 1958.
‡ ISAAK, G. R., Phys. Bull. 255, 1970.

A.2   Vector notation

A physical quantity possessing both magnitude and direction is called a vector. In figure A.2.1 the vector **A** is denoted by

$$\mathbf{A} = \mathbf{i}A_x + \mathbf{j}A_y + \mathbf{k}A_z \tag{A.2.1}$$

where **i, j, k** are vectors of unit length ('unit vectors') pointing along the

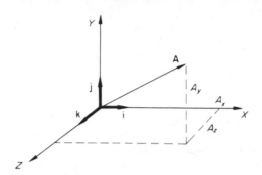

Figure A.2.1

X, Y and Z axes respectively and A_x, A_y and A_z are the projections of **A** along these axes. The 'length' of the vector is given by

$$A = |\mathbf{A}| = \sqrt{(A_x^2 + A_y^2 + A_z^2)} \tag{A.2.2}$$

Vectors add and subtract to give further vectors

$$\mathbf{A} + \mathbf{B} = \mathbf{C}$$
$$\mathbf{A} - \mathbf{B} = \mathbf{D} \tag{A.2.3}$$

where, for example

$$\mathbf{C} = \mathbf{i}(A_x + B_x) + \mathbf{j}(A_y + B_y) + \mathbf{k}(A_z + B_z)$$

A.2.1   The multiplication of vectors

The multiplication of vectors can normally be performed in two different ways; it can give the scalar (or inner) product and the vector (or outer) product. The latter is also called the skew or cross product. The scalar product of two vectors A and B is

$$\mathbf{A} \cdot \mathbf{B} = AB \cos \theta \tag{A.2.4}$$

where θ is the angle between A and B. Let us consider the above relation in terms of the unit vectors. Since **i**, **j** and **k** are mutually perpendicular

$$\mathbf{i} \cdot \mathbf{i} = \mathbf{j} \cdot \mathbf{j} = \mathbf{k} \cdot \mathbf{k} = 1$$
$$\mathbf{i} \cdot \mathbf{j} = \mathbf{i} \cdot \mathbf{k} = \mathbf{j} \cdot \mathbf{k} = \mathbf{j} \cdot \mathbf{i} = \mathbf{k} \cdot \mathbf{i} = \mathbf{k} \cdot \mathbf{j} = 0 \tag{A.2.5}$$

hence

$$\mathbf{A} \cdot \mathbf{B} = (\mathbf{i}A_x + \mathbf{j}A_y + \mathbf{k}A_z) \cdot (\mathbf{i}B_x + \mathbf{j}B_y + \mathbf{k}B_z)$$
$$= A_x B_x + A_y B_y + A_z B_z \qquad (A.2.6)$$
$$= \sum_{i=1}^{3} A_i B_i$$

where $1 \equiv x$, $2 \equiv y$, $3 \equiv z$. It is immediately obvious from the above relation that

$$\mathbf{A} \cdot \mathbf{A} = \mathbf{A}^2 = A^2 = A_x^2 + A_y^2 + A_z^2$$

which defines the length (A.2.2).

The vector product of two vectors is more complicated. Let us define a cyclic operation from figure (A.2.1)

$$\mathbf{i} \times \mathbf{j} = -\mathbf{j} \times \mathbf{i} = \mathbf{k}; \mathbf{j} \times \mathbf{k} = -\mathbf{k} \times \mathbf{j} = \mathbf{i}; \mathbf{k} \times \mathbf{i} = -\mathbf{i} \times \mathbf{k} = \mathbf{j}$$
$$\mathbf{i} \times \mathbf{i} = \mathbf{j} \times \mathbf{j} = \mathbf{k} \times \mathbf{k} = 0 \qquad (A.2.7)$$

Since $\sin \pi/2 = 1$ and $\sin 0 = 0$ the vector product must therefore involve sines (see A.2.9), furthermore

$$\mathbf{A} \times \mathbf{B} = (\mathbf{i}A_x + \mathbf{j}A_y + \mathbf{k}A_z) \times (\mathbf{i}B_x + \mathbf{j}B_y + \mathbf{k}B_z)$$
$$= \mathbf{i}(A_y B_z - A_z B_y) + \mathbf{j}(A_z B_x - A_x B_z) + \mathbf{k}(A_x B_y - A_y B_x) \qquad (A.2.8)$$
$$= \mathbf{C}$$

and we see immediately that C must be perpendicular to the plane containing A and B. If, for example, A and B lie in the YZ plane so that $A_x = B_x = 0$, then

$$\mathbf{C} = \mathbf{i}(A_y B_z - A_z B_y)$$

that is a vector pointing along the X axis. The same argument can be applied to the other planes, and so we conclude $\mathbf{C}$ is perpendicular to the plane containing $\mathbf{A}$ and $\mathbf{B}$ (figure A.2.2). The scalar product of $\mathbf{C}$ with itself is a lengthy

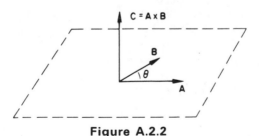

Figure A.2.2

operation, but from (A.2.6), (A.2.5) and (A.2.8) we find

$$C^2 = \mathbf{C} \cdot \mathbf{C} = (A_x^2 + A_y^2 + A_z^2)(B_x^2 + B_y^2 + B_z^2) - (A_x B_x + A_y B_y + A_z B_z)^2$$
$$= A^2 B^2 - A^2 B^2 \cos^2 \theta$$
$$= A^2 B^2 \sin^2 \theta$$

hence

$$C = |\mathbf{C}| = AB \sin\theta \qquad (A.2.9)$$

Thus **C** is a vector of length $AB \sin\theta$ in the plane perpendicular to A and B (figure A.2.2).

The relations (A.2.7) imply that $\mathbf{B} \times \mathbf{A}$ is not equal to $\mathbf{C}$; instead we find

$$\mathbf{B} \times \mathbf{A} = -\mathbf{A} \times \mathbf{B} = -\mathbf{C} \qquad (A.2.10)$$

Occasionally the triple product of vectors is encountered. We state, without proof, the following useful relationships (they can be obtained by the use of (A.2.6) and (A.2.8))

$$\mathbf{A} \cdot (\mathbf{B} \times \mathbf{C}) = \mathbf{B} \cdot (\mathbf{C} \times \mathbf{A}) = \mathbf{C} \cdot (\mathbf{A} \times \mathbf{B})$$
$$\mathbf{A} \times (\mathbf{B} \times \mathbf{C}) = -\mathbf{A} \times (\mathbf{C} \times \mathbf{B}) = (\mathbf{C} \times \mathbf{B}) \times \mathbf{A} = -(\mathbf{B} \times \mathbf{C}) \times \mathbf{A} \quad (A.2.11)$$
$$= \mathbf{B}(\mathbf{A} \cdot \mathbf{C}) - \mathbf{C}(\mathbf{A} \cdot \mathbf{B})$$

A.2.2 The differentiation of vectors
The differential operator in vector notation is*

$$\mathbf{V} = \mathbf{i}\frac{\partial}{\partial x} + \mathbf{j}\frac{\partial}{\partial y} + \mathbf{k}\frac{\partial}{\partial z} \qquad (A.2.12)$$

which is the same as (A.2.1) with the differential operator $\partial/\partial x$ replacing A_x and so on. The scalar product of **V** with **A** yields the *divergence* of **A**—this product is sometimes called div **A**.

$$\mathbf{V} \cdot \mathbf{A} = \left(\mathbf{i}\frac{\partial}{\partial x} + \mathbf{j}\frac{\partial}{\partial y} + \mathbf{k}\frac{\partial}{\partial z}\right) \cdot (\mathbf{i}A_x + \mathbf{j}A_y + \mathbf{k}A_z)$$
$$= \frac{\partial A_x}{\partial x} + \frac{\partial A_y}{\partial y} + \frac{\partial A_z}{\partial z} \qquad (A.2.13)$$

where equations (A.2.5) and (A.2.6) have been used. The vector product of **V** with **A** follows from (A.2.7) and (A.2.8)

$$\text{curl } \mathbf{A} = \mathbf{V} \times \mathbf{A} = \left(\mathbf{i}\frac{\partial}{\partial x} + \mathbf{j}\frac{\partial}{\partial y} + \mathbf{k}\frac{\partial}{\partial z}\right) \times (\mathbf{i}A_x + \mathbf{j}A_y + \mathbf{k}A_z)$$
$$= \mathbf{i}\left(\frac{\partial A_z}{\partial y} - \frac{\partial A_y}{\partial z}\right) + \mathbf{j}\left(\frac{\partial A_x}{\partial z} - \frac{\partial A_z}{\partial x}\right) + \mathbf{k}\left(\frac{\partial A_y}{\partial x} - \frac{\partial A_x}{\partial y}\right) \qquad (A.2.14)$$

Lastly the operator **V** is sometimes applied to a *scalar* (that is, a system possessing magnitude but not direction); the result is called a *gradient*

$$\text{grad } \phi = \mathbf{V}\phi = \mathbf{i}\frac{\partial\phi}{\partial x} + \mathbf{j}\frac{\partial\phi}{\partial y} + \mathbf{k}\frac{\partial\phi}{\partial z} \qquad (A.2.15)$$

---

* The symbols ∇ and ∇^2 (A.2.16) are often called 'del' and 'del squared' respectively.

where ϕ is a scalar quantity. An inspection of the above equation shows that the gradient is a vector.

A.2.3   Successive applications of ∇

The equations of the preceding sections allow us to construct the following combinations of ∇ without any great difficulty

(1) $\nabla \cdot \nabla \phi = \text{div } \textbf{grad } \phi$

$$= \left(\textbf{i} \frac{\partial}{\partial x} + \textbf{j} \frac{\partial}{\partial y} + \textbf{k} \frac{\partial}{\partial z} \right) \cdot \left(\textbf{i} \frac{\partial \phi}{\partial x} + \textbf{j} \frac{\partial \phi}{\partial y} + \textbf{k} \frac{\partial \phi}{\partial z} \right)$$

$$= \frac{\partial^2 \phi}{\partial x^2} + \frac{\partial^2 \phi}{\partial y^2} + \frac{\partial^2 \phi}{\partial z^2}$$

$$= \nabla^2 \phi$$

(A.2.16)

The symbol ∇^2 ('del squared') is sometimes also called the Laplacian operator

(2) $(\nabla \cdot \nabla) \, \textbf{A} = \nabla^2 \textbf{A} = \dfrac{\partial^2 \textbf{A}}{\partial x^2} + \dfrac{\partial^2 \textbf{A}}{\partial y^2} + \dfrac{\partial^2 \textbf{A}}{\partial z^2}$

$$= \textbf{i} \frac{\partial^2 A_x}{\partial x^2} + \textbf{j} \frac{\partial^2 A_y}{\partial y^2} + \textbf{k} \frac{\partial^2 A_z}{\partial z^2}$$

(A.2.17)

(3) $\nabla (\nabla \cdot \textbf{A}) = \textbf{grad } \text{div } \textbf{A}$

$$= \left(\textbf{i} \frac{\partial}{\partial x} + \textbf{j} \frac{\partial}{\partial y} + \textbf{k} \frac{\partial}{\partial z} \right) \left(\frac{\partial A_x}{\partial x} + \frac{\partial A_y}{\partial y} + \frac{\partial A_z}{\partial z} \right)$$

(A.2.18)

See also (A.2.21) below in connection with the above operation

(4) $\nabla \times \nabla \phi = \textbf{curl grad } \phi = 0$ (A.2.19)

(5) $\nabla \cdot \nabla \times \textbf{A} = \text{div } \textbf{curl } \textbf{A} = 0$ (A.2.20)

(6) $\nabla \times (\nabla \times \textbf{A}) = \textbf{curl curl } \textbf{A}$

$$= \nabla (\nabla \cdot \textbf{A}) - (\nabla \cdot \nabla) \, \textbf{A}$$

$$= \nabla (\nabla \cdot \textbf{A}) - \nabla^2 \textbf{A}$$

(A.2.21)

A.3   Spherical triangles and the Lorentz transformation

In the previous section the results (A.2.11)

$$\textbf{A} \cdot (\textbf{B} \times \textbf{C}) = \textbf{B} \cdot (\textbf{C} \times \textbf{A}) = \textbf{C} \cdot (\textbf{A} \times \textbf{B})$$

$$\textbf{A} \times (\textbf{B} \times \textbf{C}) = \textbf{B} (\textbf{A} \cdot \textbf{C}) - \textbf{C} (\textbf{A} \cdot \textbf{B})$$

were quoted for the triple product of vectors. Let us first extend these equations to the product of four vectors

$$(\textbf{R} \times \textbf{S}) \cdot (\textbf{T} \times \textbf{U})$$

We first write $\mathbf{P}=\mathbf{R}\times\mathbf{S}$, then from (A.2.11)

$$
\begin{aligned}
(\mathbf{R}\times\mathbf{S})\cdot(\mathbf{T}\times\mathbf{U}) &=\mathbf{P}\cdot(\mathbf{T}\times\mathbf{U})\\
&=\mathbf{T}\cdot(\mathbf{U}\times\mathbf{P})\\
&=\mathbf{T}\cdot[\mathbf{U}\times(\mathbf{R}\times\mathbf{S})]\\
&=\mathbf{T}\cdot[\mathbf{R}(\mathbf{U}\cdot\mathbf{S})-\mathbf{S}(\mathbf{U}\cdot\mathbf{R})]\\
&=(\mathbf{T}\cdot\mathbf{R})(\mathbf{U}\cdot\mathbf{S})-(\mathbf{T}\cdot\mathbf{S})(\mathbf{U}\cdot\mathbf{R})
\end{aligned}
\tag{A.3.1}
$$

We now wish to use this result in connection with figure 4.7. This diagram represented a sphere of radius im_0c; since we are primarily interested in relations between angles (equation (4.5.12) of the text) we shall consider a circle of unit radius (figure A.3.1) and multiple by im_0c later. The labels A,

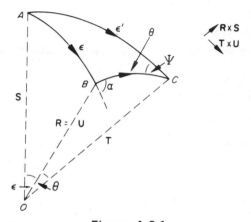

Figure A.3.1

B, C and angles in this figure are the same as in figure 4.7.

Let us return to equation (A.3.1)

$$
(\mathbf{R}\times\mathbf{S})\cdot(\mathbf{T}\times\mathbf{U})=(\mathbf{T}\cdot\mathbf{R})(\mathbf{U}\cdot\mathbf{S})-(\mathbf{T}\cdot\mathbf{S})(\mathbf{U}\cdot\mathbf{R})
$$

and consider the special case of $\mathbf{R}=\mathbf{U}$ as displayed in figure A.3.1. Since the circle is of unit radius then

$$
\begin{aligned}
(\mathbf{R}\times\mathbf{S}) &=\mathbf{e}\sin\varepsilon\\
\mathbf{R}\cdot\mathbf{S} &=\cos\varepsilon \qquad \text{and so on}
\end{aligned}
$$

where $\mathbf{e}$ is unit vector perpendicular to the plane containing $\mathbf{R}$ and $\mathbf{S}$ (compare figure A.2.2). We can now immediately interpret equation (A.3.1) as

$$
\sin\varepsilon\cos\alpha\sin\theta=\cos\theta\cos\varepsilon-\cos\varepsilon'
$$

or

$$
\cos\varepsilon'=\cos\theta\cos\varepsilon-\sin\theta\sin\varepsilon\cos\alpha
\tag{A.3.2}
$$

which is equation (4.5.12).

The above treatment also enables the first of equations (4.5.4) to be established. Since there is nothing magical about the choice of the vectors $\mathbf{R}, \mathbf{S}, \mathbf{T}, \mathbf{U}$ we may treat equation (A.3.2) in a cyclical manner (it is in fact the law of cosines for the sides of a spherical triangle). In analogy with (A.3.2) an inspection of figure A.3.1 yields

$$\cos \varepsilon = \cos \theta \, \cos \varepsilon' + \sin \theta \, \sin \varepsilon' \, \cos \psi$$

or

$$
\begin{aligned}
-\sin \varepsilon' \, \cos \psi &= \cos \varepsilon' \frac{\cos \theta}{\sin \theta} - \frac{\cos \varepsilon}{\sin \theta} \\
&= (\cos \varepsilon \, \cos \theta - \sin \theta \, \sin \varepsilon \, \cos \alpha) \frac{\cos \theta}{\sin \theta} - \frac{\cos \varepsilon}{\sin \theta} \\
&= \frac{\cos \varepsilon}{\sin \theta} (\cos^2 \theta - 1) - \sin \varepsilon \, \cos \alpha \, \cos \theta \\
&= -\cos \varepsilon \, \sin \theta - \sin \varepsilon \, \cos \alpha \, \cos \theta
\end{aligned}
$$

where we have used (A.3.2). If this equation is multiplied through by $im_0 c$ we find

$$p' \cos \psi = E\beta \gamma + p \cos \alpha \gamma \qquad (A.3.3)$$

If we use the direction $\boldsymbol{\beta}$ as axis of orientation then

$$p' \cos \psi = \mathbf{p}'_{\parallel} \qquad p \cos \alpha = \mathbf{p}_{\parallel}$$

and so equation (A.3.3) becomes

$$\mathbf{p}'_{\parallel} = \gamma \left(\mathbf{p}_{\parallel} + \beta E \right)$$

which is the first of equations (4.5.4). The second of these equations

$$\mathbf{p}'_{\perp} = \mathbf{p}_{\perp}$$

implies that

$$\sin \varepsilon' \, \sin \psi = \sin \varepsilon \, \sin (\pi - \alpha)$$

which is easily recognisable as the standard law of sines

$$\frac{\sin \varepsilon'}{\sin (\pi - \alpha)} = \frac{\sin \varepsilon}{\sin \psi}$$

for spherical triangles.

It is apparent from the above relations that the methodology of spherical triangles offers a fairly simple technique for handling successive Lorentz transformations.

Solutions to problems

1.1 (a) 62 mile/h; (b) 58 mile/h.

1.2 1.6*c*.

1.4 Velocity of a = velocity of b = $\dfrac{10}{\sqrt{2}}$ units

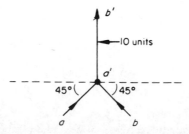

where the broken line represents the original direction of a.

1.6 $\delta t = v^2 l \cos 2\theta / c^3$.

158

2.3 1, 1.02, 1.09, 1.25, 1.67, 2.29, 7.09.

2.4 $x'_1 = 17.5$ m, $t'_1 = 1.25 \times 10^{-7}$ s, $t'_2 - t'_1 = 2.25 \times 10^{-0}$ s.

2.5 $x = \pm 100$ m.

2.8 Time-like.

3.1 0, 450, 4200 m; 0.360, 594 m.

3.7 Jan. 1, 1990; Jan. 1, 1986.

3.8 A Oct. 1981; B Mar. 1988.

3.10 $0.975c$.

4.2 (a) 3.57×10^{-13} joule; (b) 2.22 MeV.

4.3 0.938 GeV/c^2; 1.007 a.m.u.; 1.13 GeV; 0.38.

4.6 0.00335; 0.55.

4.7 $\tan \theta' = \dfrac{m_d c^2 p_a \sin \theta}{\omega_d p_a \cos - \omega_a p_d}$

ω, p are energy and linear momenta ($|\mathbf{p}|$) respectively.

4.8 1230 GeV.

4.9 $\gamma_2(\gamma_1 \beta_1 p_x + \beta_2 p_y) c + \gamma_1 \gamma_2 E$

$\gamma_1(\beta_1 p_x + \gamma_2 \beta_2 p_y) c + \gamma_1 \gamma_2 E$

4.10 6.9 GeV/c.

4.11 $0.51c$.

4.12 21.4 m; no.

5.2 $53° 30'$, $43° 6'$; $22° 30'$, $43° 6'$.

5.5 $\tan^{-1} \dfrac{1}{\gamma_\pi(L)} \tan \alpha$.

5.8 7.5×10^{-35}.

6.12 17.

6.13 $\mathbf{u}$ remains constant; spin precesses about $\mathbf{B}$ with frequency $geB/2m_0\gamma$.

Index